On the Existence
of Feller Semigroups
with Boundary Conditions

Recent Titles in This Series

(*Continued in the back of this publication*)

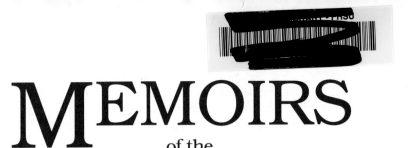

MEMOIRS
of the
American Mathematical Society

Number 475

On the Existence
of Feller Semigroups
with Boundary Conditions

Kazuaki Taira

September 1992 • Volume 99 • Number 475 (second of 4 numbers) • ISSN 0065-9266

American Mathematical Society
Providence, Rhode Island

1991 *Mathematics Subject Classification.*
Primary 47D07, 35J25; Secondary 47D05, 60J35, 60J60.

Library of Congress Cataloging-in-Publication Data

Taira, Kazuaki.
  On the existence of Feller semigroups with boundary conditions/Kazuaki Taira.
  p. cm. – (Memoirs of the American Mathematical Society, ISSN 0065-9266; no. 475)
  "September 1992, volume 99, number 475 (second of 4 numbers)."
  Includes bibliographical references (p.).
  ISBN 0-8218-2535-6
  1. Markov processes.   2. Differential equations, Elliptic.   3. Boundary value problems.   I. Title.
  II. Title: Feller semigroups with boundary conditions.   III. Series.
  QA3.A57 no. 475
  [QA274.7]
  510 s–dc20   92-18061
  [519.2'33]   CIP

Memoirs of the American Mathematical Society
This journal is devoted entirely to research in pure and applied mathematics.

Subscription information. The 1992 subscription begins with Number 459 and consists of six mailings, each containing one or more numbers. Subscription prices for 1992 are $292 list, $234 institutional member. A late charge of 10% of the subscription price will be imposed on orders received from nonmembers after January 1 of the subscription year. Subscribers outside the United States and India must pay a postage surcharge of $25; subscribers in India must pay a postage surcharge of $43. Expedited delivery to destinations in North America $30; elsewhere $82. Each number may be ordered separately; *please specify number* when ordering an individual number. For prices and titles of recently released numbers, see the New Publications sections of the *Notices of the American Mathematical Society.*
  Back number information. For back issues see the *AMS Catalog of Publications.*
  Subscriptions and orders should be addressed to the American Mathematical Society, P. O. Box 1571, Annex Station, Providence, RI 02901-1571. *All orders must be accompanied by payment.* Other correspondence should be addressed to Box 6248, Providence, RI 02940-6248.

Memoirs of the American Mathematical Society is published bimonthly (each volume consisting usually of more than one number) by the American Mathematical Society at 201 Charles Street, Providence, RI 02904-2213. Second-class postage paid at Providence, Rhode Island. Postmaster: Send address changes to Memoirs, American Mathematical Society, P. O. Box 6248, Providence, RI 02940-6248.

TABLE OF CONTENTS

ABSTRACT

This paper is devoted to the functional analytic approach to the problem of construction of Feller semigroups with Ventcel' (Wentzell) boundary conditions. The problem of construction of Feller semigroups has never before been studied in the *non-transversal* case. In this paper we consider the non-transversal case, and solve from the viewpoint of functional analysis the problem of construction of Feller semigroups for *elliptic* Waldenfels operators. Intuitively, our result may be stated as follows : One can construct a Feller semigroup corresponding to such a diffusion phenomenon that a Markovian particle moves both by jumps and continuously in the state space until it "dies" at which time it reaches the set where the absorption phenomenon occurs.

1991 *Mathematics Subject Classification.* Primary 47D07, 35J25 ; Secondary 47D05, 60J35, 60J60.

Key words and phrases : Feller semigroups, Markov processes, elliptic boundary value problems.

This research was partially supported by Grant-in-Aid for General Scientific Research (No. 03640122), Ministry of Education, Science and Culture.

ACKNOWLEDGMENTS

A major part of the work was done at Ecole Polytechnique (Palaiseau, France) in May 1990 while the author was on leave from University of Tsukuba. The author would like to thank Professor Jean-Michel Bony for valuable discussions. Thanks are also due to Ecole Polytechnique for its hospitality.

Dedicated to the memory of my father, Yasunori TAIRA (1915-1990)

INTRODUCTION AND RESULTS

Let D be a bounded domain of Euclidean space $\mathbf{R}^N$, with C^∞ boundary ∂D; its closure $\bar{D} = D \cup \partial D$ is an N-dimensional, compact C^∞ manifold with boundary.

Let $C(\bar{D})$ be the space of real-valued, continuous functions on $\bar{D}$. We equip the space $C(\bar{D})$ with the topology of uniform convergence on the whole $\bar{D}$; hence it is a Banach space with the maximum norm

$$\|f\| = \max_{x \in \bar{D}} |f(x)|.$$

A strongly continuous semigroup $\{T_t\}_{t \geq 0}$ on the space $C(\bar{D})$ is called a *Feller semigroup* on $\bar{D}$ if it is non-negative and contractive on $C(\bar{D})$:

$$f \in C(\bar{D}), \ 0 \leq f \leq 1 \text{ on } \bar{D} \implies 0 \leq T_t f \leq 1 \text{ on } \bar{D}.$$

It is known (cf. [12]) that if T_t is a Feller semigroup on $\bar{D}$, then there exists a unique Markov transition function p_t on $\bar{D}$ such that

$$T_t f(x) = \int_{\bar{D}} p_t(x, dy) f(y), \quad f \in C(\bar{D}).$$

It can be shown that the function p_t is the transition function of some strong *Markov process*; hence the value $p_t(x, E)$ expresses the transition probability that a Markovian particle starting at position x will be found in the set E at time t.

Furthermore, it is known (cf. [1], [9], [12], [17]) that the infinitesimal generator $\mathfrak{A}$ of a Feller semigroup $\{T_t\}_{t \geq 0}$ is described analytically by a Waldenfels operator W and a Ventcel' boundary condition L, which we formulate precisely.

Let W be a second-order *elliptic* integro-differential operator with real coefficients such that

$$(0.1) \quad Wu(x) = Pu(x) + S_r u(x)$$

$$\equiv \left(\sum_{i,j=1}^N a^{ij}(x) \frac{\partial^2 u}{\partial x_i \partial x_j}(x) + \sum_{i=1}^N b^i(x) \frac{\partial u}{\partial x_i}(x) + c(x)u(x) \right)$$

$$+ \left(\int_D s(x,y) \left[u(y) - \sigma(x,y) \left(u(x) + \sum_{j=1}^N (y_j - x_j) \frac{\partial u}{\partial x_j}(x) \right) \right] dy \right),$$

where:

1) $a^{ij} \in C^\infty(\mathbf{R}^N)$, $a^{ij} = a^{ji}$ and there exists a constant $a_0 > 0$ such that

$$\sum_{i,j=1}^N a^{ij}(x) \xi_i \xi_j \geq a_0 |\xi|^2, \quad x \in \mathbf{R}^N, \ \xi \in \mathbf{R}^N.$$

Received by the editor November 8, 1990.

2) $b^i \in C^\infty(\mathbf{R}^N)$.

3) $c \in C^\infty(\mathbf{R}^N)$ and $c \leq 0$ in D.

4) The integral kernel $s(x, y)$ is the distribution kernel of a properly supported, pseudo-differential operator $S \in L_{1,0}^{2-\kappa}(\mathbf{R}^N)$, $\kappa > 0$, which has the *transmission property* with respect to the boundary ∂D (cf. Section 2.2), and $s(x, y) \geq 0$ off the diagonal $\{(x, x); x \in \mathbf{R}^N\}$ in $\mathbf{R}^N \times \mathbf{R}^N$. The measure dy is the Lebesgue measure on $\mathbf{R}^N$.

5) The function $\sigma(x, y)$ is a C^∞ function on $\bar{D} \times \bar{D}$ such that $\sigma(x, y) = 1$ in a neighborhood of the diagonal $\{(x, x); x \in \bar{D}\}$ in $\bar{D} \times \bar{D}$. The function $\sigma(x, y)$ depends on the shape of the domain D. More precisely, it depends on a family of local charts on D in each of which the Taylor expansion is valid for functions u. For example, if D is convex, one may take $\sigma(x, y) \equiv 1$ on $\bar{D} \times \bar{D}$.

6) $W1(x) = c(x) + \int_D s(x, y)[1 - \sigma(x, y)]dy \leq 0$ in D.

The operator W is called a second-order *Waldenfels operator* (cf. [1]). The differential operator P is called a diffusion operator which describes analytically a strong Markov process with continuous paths (diffusion process) in the interior D. The operator S_r is called a second-order Lévy operator which is supposed to correspond to the jump phenomenon in the interior D; a Markovian particle moves by jumps to a random point, chosen with kernel $s(x, y)$ and function $\sigma(x, y)$, in the interior D. Therefore, the Waldenfels operator W is supposed to correspond to such a diffusion phenomenon that a Markovian particle moves both by jumps and continuously in the state space D.

The intuitive meaning of condition 6) is that the jump phenomenon from a point $x \in D$ to the outside of a neighborhood of x in D is "dominated" by the absorption phenomenon at x. We remark that in the case when $\sigma(x, y) \equiv 1$ on $\bar{D} \times \bar{D}$, condition 6) is reduced to the following simple one:

6') $W1(x) = c(x) \leq 0$ in D.

Let L be a second-order boundary condition such that in local coordinates $(x_1, \cdots, x_{N-1})$

(0.2)
$$
Lu(x') = Qu(x') + \mu(x')\frac{\partial u}{\partial \mathbf{n}}(x') - \delta(x')Wu(x') + \Gamma u(x')
$$
$$
\equiv \left(\sum_{i,j=1}^{N-1} \alpha^{ij}(x')\frac{\partial^2 u}{\partial x_i \partial x_j}(x') + \sum_{i=1}^{N-1} \beta^i(x')\frac{\partial u}{\partial x_i}(x') + \gamma(x')u(x') \right)
$$
$$
+ \mu(x')\frac{\partial u}{\partial \mathbf{n}}(x') - \delta(x')Wu(x') + \left(\eta(x')u(x') + \sum_{i=1}^{N-1} \zeta^i(x')\frac{\partial u}{\partial x_i}(x') \right.
$$
$$
+ \int_{\partial D} r(x', y')\left[u(y') - \tau(x', y')\left(u(x') + \sum_{j=1}^{N-1} (y_j - x_j)\frac{\partial u}{\partial x_j}(x') \right) \right]dy'
$$
$$
\left. + \int_D t(x', y)\left[u(y) - \tau(x', y)\left(u(x') + \sum_{j=1}^{N-1} (y_j - x_j)\frac{\partial u}{\partial x_j}(x') \right) \right]dy \right),
$$

where:

1) The operator Q is a second-order degenerate elliptic differential operator on ∂D with non-positive principal symbol. In other words, the α^{ij} are the components of a C^∞ symmetric contravariant tensor of type $\binom{2}{0}$ on ∂D satisfying

$$\sum_{i,j=1}^{N-1} \alpha^{ij}(x')\xi_i\xi_j \geq 0, \ \ x' \in \partial D, \ \xi' = \sum_{j=1}^{N-1} \xi_j\, dx_j \in T_{x'}^*(\partial D).$$

Here $T_{x'}^*(\partial D)$ is the cotangent space of ∂D at x'.

2) $Q1 = \gamma \in C^\infty(\partial D)$ and $\gamma \leq 0$ on ∂D.

3) $\mu \in C^\infty(\partial D)$ and $\mu \geq 0$ on ∂D.

4) $\delta \in C^\infty(\partial D)$ and $\delta \geq 0$ on ∂D.

5) $\mathbf{n} = (n_1, \ldots, n_N)$ is the unit interior normal to the boundary ∂D.

6) The integral kernel $r(x', y')$ is the distribution kernel of a pseudo-differential operator $R \in L_{1,0}^{2-\kappa_1}(\partial D)$, $\kappa_1 > 0$, and $r(x', y') \geq 0$ off the diagonal $\Delta_{\partial D} = \{(x', x'); x' \in \partial D\}$ in $\partial D \times \partial D$. The density dy' is a strictly positive density on ∂D.

7) The integral kernel $t(x, y)$ is the distribution kernel of a properly supported, pseudo-differential operator $T \in L_{1,0}^{2-\kappa_2}(\mathbf{R}^N)$, $\kappa_2 > 0$, which has the transmission property with respect to the boundary ∂D, and $t(x, y) \geq 0$ off the diagonal $\{(x, x); x \in \mathbf{R}^N\}$ in $\mathbf{R}^N \times \mathbf{R}^N$.

8) The function $\tau(x, y)$ is a C^∞ function on $\bar{D} \times \bar{D}$, with compact support in a neighborhood of the diagonal $\Delta_{\partial D}$, such that, at each point x' of ∂D, $\tau(x', y) = 1$ for y in a neighborhood of x' in $\bar{D}$. The function $\tau(x, y)$ depends on the shape of the boundary ∂D.

9) The operator Γ is a boundary condition of order $2 - \min(\kappa_1, \kappa_2)$, and satisfies the condition

$$\Gamma 1(x') = \eta(x') + \int_{\partial D} r(x', y')[1 - \tau(x', y')]dy'$$
$$+ \int_D t(x', y)[1 - \tau(x', y)]dy \leq 0 \text{ on } \partial D.$$

The boundary condition L is called a second-order *Ventcel' boundary condition* (cf. [17]). The terms of L

$$\sum_{i,j=1}^{N-1} \alpha^{ij}(x')\frac{\partial^2 u}{\partial x_i \partial x_j}(x') + \sum_{i=1}^{N-1} \left(\beta^i(x') + \zeta^i(x')\right)\frac{\partial u}{\partial x_i}(x'),$$

$$(\gamma(x') + \eta(x'))\, u(x'), \ \mu(x')\frac{\partial u}{\partial \mathbf{n}}(x'), \ \delta(x')Wu(x'),$$

$$\int_{\partial D} r(x', y')\left[u(y') - \tau(x', y')\left(u(x') + \sum_{j=1}^{N-1}(y_j - x_j)\frac{\partial u}{\partial x_j}(x')\right)\right]dy',$$

$$\int_D t(x', y)\left[u(y) - \tau(x', y)\left(u(x') + \sum_{j=1}^{N-1}(y_j - x_j)\frac{\partial u}{\partial x_j}(x')\right)\right]dy$$

are supposed to correspond to the diffusion along the boundary, the absorption phenomenon, the reflection phenomenon, the viscosity phenomenon and the jump phenomenon on the boundary and the inward jump phenomenon from the boundary, respectively.

The intuitive meaning of condition 9) is that the jump phenomenon from a point $x' \in \partial D$ to the outside of a neighborhood of x' in $\bar{D}$ is "dominated" by the absorption phenomenon at x'.

This paper is devoted to the functional analytic approach to the problem of construction of Feller semigroups with Ventcel' boundary conditions. More precisely, we consider the following problem:

Problem. *Conversely, given analytic data* (W, L), *can we construct a Feller semigroup* $\{T_t\}_{t \geq 0}$ *whose infinitesimal generator* $\mathfrak{A}$ *is characterized by* (W, L) ?

We say that the boundary condition L is *transversal* on the boundary ∂D if it satisfies the condition:

$$\int_D t(x', y) dy = +\infty \quad \text{if } \mu(x') = \delta(x') = 0.$$

Intuitively, the transversality condition implies that a Markovian particle jumps away "instantaneously" from the points $x' \in \partial D$ where neither reflection nor viscosity phenomenon occurs (which is similar to the reflection phenomenon). Probabilistically, this means that every Markov process on the boundary ∂D is the "trace" on ∂D of trajectories of some Markov process on the closure $\bar{D} = D \cup \partial D$.

The next theorem asserts that there exists a Feller semigroup on $\bar{D}$ corresponding to such a diffusion phenomenon that one of the reflection phenomenon, the viscosity phenomenon and the inward jump phenomenon from the boundary occurs at each point of the boundary ∂D:

Theorem 1. *We define a linear operator* $\mathfrak{A}$ *from the space* $C(\bar{D})$ *into itself as follows:*

(a) The domain of definition $D(\mathfrak{A})$ *of* $\mathfrak{A}$ *is the set*

$$(0.3) \qquad D(\mathfrak{A}) = \left\{ u \in C(\bar{D}); Wu \in C(\bar{D}), Lu = 0 \right\}.$$

(b) $\mathfrak{A}u = Wu, \ u \in D(\mathfrak{A})$.
Here Wu *and* Lu *are taken in the sense of distributions.*

Assume that the boundary condition L *is* transversal *on the boundary* ∂D. *Then the operator* $\mathfrak{A}$ *generates a Feller semigroup* $\{T_t\}_{t \geq 0}$ *on* $\bar{D}$.

We remark that Theorem 1 was proved before by Taira [12] under some additional conditions (cf. [12, Theorem 10.1.3]), and also by Cancelier [3] (cf. [3, Théorème 3.2]). Takanobu and Watanabe [14] proved a probabilistic version of Theorem 1 in the case when the domain D is the half space $\mathbf{R}_+^N$ (cf. [14, Corollary]).

The problem of construction of Feller semigroups has never before, to the author's knowledge, been studied in the *non-transversal* case. In this paper we consider the following case:

(A) $\mu(x') + \delta(x') - \gamma(x') > 0$ on ∂D.

Intuitively, condition (A) implies that one of the absorption, reflection, viscosity phenomena occurs at each point of the boundary ∂D.

Furthermore, we assume that:

(H) There exists a second-order Ventcel' boundary condition L_ν such that

$$Lu(x') = \mu(x')L_\nu u(x') + \gamma(x')u(x') - \delta(x')Wu(x'), \quad x' \in \partial D,$$

where the boundary condition L_ν is given in local coordinates $(x_1, \cdots, x_{N-1})$ by the formula

$$
\begin{aligned}
L_\nu u(x') = &\sum_{i,j=1}^{N-1} \bar{\alpha}^{ij}(x')\frac{\partial^2 u}{\partial x_i \partial x_j}(x') + \sum_{i=1}^{N-1} \bar{\beta}^i(x')\frac{\partial u}{\partial x_i}(x')\\
&+ \frac{\partial u}{\partial \mathbf{n}}(x') + \bar{\eta}(x')u(x') + \sum_{i=1}^{N-1} \bar{\zeta}^i(x')\frac{\partial u}{\partial x_i}(x')\\
&+ \int_{\partial D} \bar{r}(x',y')\left[u(y') - \tau(x',y')\left(u(x') + \sum_{j=1}^{N-1}(y_j - x_j)\frac{\partial u}{\partial x_j}(x')\right)\right]dy'\\
&+ \int_{D} \bar{t}(x',y)\left[u(y) - \tau(x',y)\left(u(x') + \sum_{j=1}^{N-1}(y_j - x_j)\frac{\partial u}{\partial x_j}(x')\right)\right]dy,
\end{aligned}
$$

and satisfies the condition

$$\bar{\eta}(x') + \int_{\partial D} \bar{r}(x',y')[1 - \tau(x',y')]dy' + \int_{D} \bar{t}(x',y)[1 - \tau(x',y)]dy \le 0 \text{ on } \partial D.$$

We remark that the boundary condition L is *not* transversal on ∂D, while the boundary condition L_ν is transversal on ∂D, since $\mu(x') \equiv 1$ on ∂D.

Intuitively, condition (H) implies that the diffusion along the boundary, the inward jump phenomenon from the boundary and the jump phenomenon on the boundary are "dominated" by the reflection phenomenon.

Now we introduce a subspace of $C(\bar{D})$ which is associated with the boundary condition L.

We let

$$M = \{x' \in \partial D; \mu(x') = \delta(x') = 0, \int_{D} t(x',y)dy < \infty\}.$$

Then, by condition (H), we find that

$$M = \{x' \in \partial D; \mu(x') = \delta(x') = 0\}.$$

Further, in view of condition (A), it follows that the boundary condition

$$Lu = \mu L_\nu u + \gamma(u|_{\partial D}) - \delta(Wu|_{\partial D}) = 0 \text{ on } \partial D$$

includes the condition

$$u = 0 \text{ on } M.$$

With this fact in mind, we let

$$C_0(\bar{D}\backslash M) = \{u \in C(\bar{D}); u = 0 \text{ on } M\}.$$

The space $C_0(\bar{D}\backslash M)$ is a closed subspace of $C(\bar{D})$; hence it is a Banach space.

A strongly continuous semigroup $\{T_t\}_{t\geq 0}$ on the space $C_0(\bar{D}\backslash M)$ is called a *Feller semigroup* on $\bar{D}\backslash M$ if it is non-negative and contractive on $C_0(\bar{D}\backslash M)$:

$$f \in C_0(\bar{D}\backslash M), 0 \leq f \leq 1 \text{ on } \bar{D}\backslash M \implies 0 \leq T_t f \leq 1 \text{ on } \bar{D}\backslash M.$$

We define a linear operator $\mathfrak{A}$ from $C_0(\bar{D}\backslash M)$ into itself as follows:

(a) The domain of definition $D(\mathfrak{A})$ of $\mathfrak{A}$ is the set

(0.4) $$D(\mathfrak{A}) = \left\{u \in C_0(\bar{D}\backslash M); Wu \in C_0(\bar{D}\backslash M), Lu = 0\right\}.$$

(b) $\mathfrak{A}u = Wu$, $u \in D(\mathfrak{A})$.

The next theorem is a generalization of Theorem 1 to the non-transversal case:

Theorem 2. *Assume that the following conditions (A) and (H) are satisfied:*

(A) $\mu(x') + \delta(x') - \gamma(x') > 0$ *on* ∂D.

(H) *There exists a transversal Ventcel' boundary condition L_ν of second order such that*

$$Lu(x') = \mu(x')L_\nu u(x') + \gamma(x')u(x') - \delta(x')Wu(x'), \quad x' \in \partial D.$$

Then the operator $\mathfrak{A}$ defined by formula (0.4) generates a Feller semigroup $\{T_t\}_{t\geq 0}$ on $\bar{D}\backslash M$.

If T_t is a Feller semigroup on $\bar{D}\backslash M$, then there exists a unique Markov transition function p_t on $\bar{D}\backslash M$ such that

$$T_t f(x) = \int_{\bar{D}\backslash M} p_t(x, dy) f(y), \quad f \in C_0(\bar{D}\backslash M),$$

and further p_t is the transition function of some strong Markov process. On the other hand, the intuitive meaning of conditions (A) and (H) is that the absorption phenomenon occurs at each point of the set $M = \{x' \in \partial D; \mu(x') = \delta(x') = 0\}$. Therefore, Theorem 2 asserts that there exists a Feller semigroup on $\bar{D}\backslash M$ corresponding to such a diffusion phenomenon that a Markovian particle moves both by jumps and continuously in the state space $\bar{D}\backslash M$ until it "dies" at which time it reaches the set M.

We remark that Taira [13] has proved Theorem 2 under the condition that $L_\nu = \partial/\partial\mathbf{n}$ and $\delta \equiv 0$ on ∂D, by using the L^p theory of pseudo-differential operators (cf. [13, Theorem 4]).

Finally we consider the case when all the operators S, T and R are pseudo-differential operators of order *less than one*. Then one can take $\sigma(x, y) \equiv 1$ on $\bar{D} \times \bar{D}$, and write the operator W in the following form:

(0.1′) $$Wu(x) = Pu(x) + S_r u(x)$$

$$\equiv \left(\sum_{i,j=1}^{N} a^{ij}(x)\frac{\partial^2 u}{\partial x_i \partial x_j}(x) + \sum_{i=1}^{N} b^i(x)\frac{\partial u}{\partial x_i}(x) + c(x)u(x)\right)$$

$$+ \left(\int_D s(x, y)[u(y) - u(x)]dy\right),$$

where:

4') The integral kernel $s(x, y)$ is the distribution kernel of a properly supported, pseudo-differential operator $S \in L_{1,0}^{1-\kappa}(\mathbf{R}^N)$, $\kappa > 0$, which has the transmission property with respect to the boundary ∂D, and $s(x, y) \geq 0$ off the diagonal $\{(x, x); x \in \mathbf{R}^N\}$ in $\mathbf{R}^N \times \mathbf{R}^N$.

6') $W1(x) = c(x) \leq 0$ in D.

Similarly, the boundary condition L can be written in the following form:

$$(0.2') \quad Lu(x') = Qu(x') + \mu(x')\frac{\partial u}{\partial \mathbf{n}}(x') - \delta(x')Wu(x') + \Gamma u(x')$$

$$\equiv \left(\sum_{i,j=1}^{N-1} \alpha^{ij}(x')\frac{\partial^2 u}{\partial x_i \partial x_j}(x') + \sum_{i=1}^{N-1} \beta^i(x')\frac{\partial u}{\partial x_i}(x') + \gamma(x')u(x') \right)$$

$$+ \mu(x')\frac{\partial u}{\partial \mathbf{n}}(x') - \delta(x')Wu(x') + \left(\eta(x')u(x') + \sum_{i=1}^{N-1} \zeta^i(x')\frac{\partial u}{\partial x_i}(x') \right.$$

$$\left. + \int_{\partial D} r(x', y')[u(y') - u(x')]dy' + \int_D t(x', y)[u(y) - u(x')]dy \right),$$

where:

6') The integral kernel $r(x', y')$ is the distribution kernel of a pseudo-differential operator $R \in L_{1,0}^{1-\kappa_1}(\partial D)$, $\kappa_1 > 0$, and $r(x', y') \geq 0$ off the diagonal $\{(x', x'); x' \in \partial D\}$ in $\partial D \times \partial D$.

7') The integral kernel $t(x, y)$ is the distribution kernel of a properly supported, pseudo-differential operator $T \in L_{1,0}^{1-\kappa_2}(\mathbf{R}^N)$, $\kappa_2 > 0$, which has the transmission property with respect to the boundary ∂D, and $t(x, y) \geq 0$ off the diagonal $\{(x, x); x \in \mathbf{R}^N\}$ in $\mathbf{R}^N \times \mathbf{R}^N$.

9') $\Gamma 1(x') = \eta(x') \leq 0$ on ∂D.

Then Theorems 1 and 2 may be simplified as follows:

Theorem 3. *Assume that the operator W and the boundary condition L are of the forms (0.1') and (0.2'), respectively. If the boundary condition L is transversal on the boundary ∂D, then the operator $\mathfrak{A}$ defined by formula (0.3) generates a Feller semigroup $\{T_t\}_{t \geq 0}$ on $\bar{D}$.*

Theorem 4. *Assume that the operator W and the boundary condition L are of the forms (0.1') and (0.2'), respectively. If conditions (A) and (H) are satisfied, then the operator $\mathfrak{A}$ defined by formula (0.4) generates a Feller semigroup $\{T_t\}_{t \geq 0}$ on $\bar{D} \backslash M$.*

Theorems 1, 2, 3 and 4 solve from the viewpoint of functional analysis the problem of construction of Feller semigroups with Ventcel' boundary conditions for *elliptic* Waldenfels operators.

The rest of this paper is organized as follows.

In Chapter I, we present a brief description of the basic definitions and results about a class of semigroups (Feller semigroups) associated with Markov processes, which forms a functional analytic background for the proof of Theorems 1 and 2.

Chapter II provides a review of the basic concepts and results of the theory of pseudo-differential operators - a modern theory of potentials - which will be used in subsequent chapters. In particular, we introduce the notion of transmission property, due to Boutet de Monvel [2], which is a condition about symbols in the normal direction at the boundary. Furthermore, we prove an existence and uniqueness theorem for a class of pseudo-differential operators which enters naturally in the construction of Feller semigroups.

Chapter III is devoted to the proof of Theorem 1. We reduce the problem of construction of Feller semigroups to the problem of *unique solvability* for the boundary value problem

$$\begin{cases} (\alpha - W)u = f & \text{in } D, \\ (\lambda - L)u = \varphi & \text{on } \partial D, \end{cases}$$

and then prove existence theorems for Feller semigroups. Here $\alpha > 0$ and $\lambda \geq 0$.

The idea of our approach is stated as follows (cf. [1], [9], [12]).

First we consider the following *Dirichlet problem*:

$$\begin{cases} (\alpha - W)v = f & \text{in } D, \\ v|_{\partial D} = 0 & \text{on } \partial D. \end{cases}$$

The existence and uniqueness theorem for this problem is well established in the framework of Hölder spaces. We let

$$v = G_\alpha^0 f.$$

The operator G_α^0 is the Green operator for the Dirichlet problem. Then it follows that a function u is a solution of the problem

$$(*) \qquad\qquad \begin{cases} (\alpha - W)u = f & \text{in } D, \\ Lu = 0 & \text{on } \partial D \end{cases}$$

if and only if the function $w = u - v$ is a solution of the problem

$$\begin{cases} (\alpha - W)w = 0 & \text{in } D, \\ Lw = -Lv = -LG_\alpha^0 f & \text{on } \partial D. \end{cases}$$

But we know that every solution w of the equation

$$(\alpha - W)w = 0 \ \text{ in } D$$

can be expressed by means of a single layer potential as follows:

$$w = H_\alpha \psi.$$

The operator H_α is the harmonic operator for the Dirichlet problem. Thus, by using the Green and harmonic operators, one can reduce the study of problem $(*)$ to that of the equation:

$$LH_\alpha \psi = -LG_\alpha^0 f.$$

This is a generalization of the classical Fredholm integral equation. It is known (cf. [2], [5], [10]) that the operator LH_α is a pseudo-differential operator of second order on the boundary ∂D.

By using the Hölder space theory of pseudo-differential operators, we can show that if the boundary condition L is transversal on the boundary ∂D, then the operator LH_α is *bijective* in the framework of Hölder spaces. The crucial point in the proof is that we consider the term $\delta(Wu|_{\partial D})$ of viscosity in the boundary condition

$$Lu = L_0 u - \delta(Wu|_{\partial D})$$

as a term of "perturbation" of the boundary condition $L_0 u$.

Therefore, we find that a unique solution u of problem $(*)$ can be expressed as follows:

$$u = G_\alpha^0 f - H_\alpha \left(LH_\alpha^{-1} LG_\alpha^0 f \right).$$

This formula allows us to verify all the conditions of the generation theorems of Feller semigroups discussed in Chapter I. Intuitively, this formula tells us that if the boundary condition L is transversal on the boundary ∂D, then one can "piece together" a Markov process on the boundary ∂D with W-diffusion in the interior D to construct a Markov process on the closure $\bar{D} = D \cup \partial D$.

In Chapter IV, we prove Theorem 2. We explain the idea of the proof.

First we remark that if condition (H) is satisfied, then the boundary condition L can be written in the following form:

$$Lu = \mu L_\nu u + \gamma(u|_{\partial D}) - \delta(Wu|_{\partial D}),$$

where the boundary condition L_ν is *transversal* on ∂D. Hence, applying Theorem 1 to the boundary condition L_ν, we can solve uniquely the following boundary value problem:

$$\begin{cases} (\alpha - W)v = f & \text{in } D, \\ L_\nu v = 0 & \text{on } \partial D. \end{cases}$$

We let

$$v = G_\alpha^\nu f.$$

The operator G_α^ν is the Green operator for the boundary condition L_ν. Then it follows that a function u is a solution of the problem

$(**)$
$$\begin{cases} (\alpha - W)u = f & \text{in } D, \\ Lu = \mu L_\nu u + \gamma(u|_{\partial D}) - \delta(Wu|_{\partial D}) = 0 & \text{on } \partial D \end{cases}$$

if and only if the function $w = u - v$ is a solution of the problem

$$\begin{cases} (\alpha - W)w = 0 & \text{in } D, \\ Lw = -Lv = (\alpha\delta - \gamma)(v|_{\partial D}) - \delta(f|_{\partial D}) & \text{on } \partial D. \end{cases}$$

Thus, as in the proof of Theorem 1, one can reduce the study of problem $(**)$ to that of the equation:

$$LH_\alpha \psi = (\alpha\delta - \gamma)(G_\alpha^\nu f|_{\partial D}) - \delta(f|_{\partial D}).$$

By using the Hölder space theory of pseudo-differential operators as in the proof of Theorem 1, we can show that if condition (A) is satisfied, then the operator LH_α is *bijective* in the framework of Hölder spaces.

Therefore, we find that a unique solution u of problem $(**)$ can be expressed as follows:

$$u = G_\alpha^\nu f - H_\alpha \left(LH_\alpha^{-1} \, LG_\alpha^\nu f \right).$$

This formula allows us to verify all the conditions of the generation theorems of Feller semigroups, especially the *density* of the domain $D(\mathfrak{A})$ in the space $C_0(\bar{D}\backslash M)$.

If we use instead of G_α^ν the Green operator G_α^0 for the Dirichlet problem as in the proof of Theorem 1, our proof would break down.

We do not prove Theorems 3 and 4, since their proofs are essentially the same as those of Theorems 1 and 2, respectively.

CHAPTER I

THEORY OF FELLER SEMIGROUPS

This chapter provides a brief description of the basic definitions and results about a class of semigroups (Feller semigroups) associated with Markov processes, which forms a functional analytic background for the proof of Theorems 1 and 2. The results discussed here are adapted from Chapter 9 of Taira [12].

1.1 Markov Transition Functions and Feller Semigroups

Let (K, ρ) be a locally compact, separable metric space and $\mathcal{B}$ the σ-algebra of all Borel sets in K.

A function $p_t(x, E)$, defined for all $t \geq 0$, $x \in K$ and $E \in \mathcal{B}$, is called a (temporally homogeneous) *Markov transition function* on K if it satisfies the following four conditions:

(a) $p_t(x, \cdot)$ is a non-negative measure on $\mathcal{B}$ and $p_t(x, K) \leq 1$ for each $t \geq 0$ and each $x \in K$.

(b) $p_t(\cdot, E)$ is a Borel measurable function for each $t \geq 0$ and each $E \in \mathcal{B}$.

(c) $p_0(x, \{x\}) = 1$ for each $x \in K$.

(d) (The Chapman-Kolmogorov equation) For any t, $s \geq 0$, $x \in K$ and any $E \in \mathcal{B}$, we have

$$(1.1) \qquad p_{t+s}(x, E) = \int_K p_t(x, dy) p_s(y, E).$$

Here is an intuitive way of thinking about the above definition of a Markov transition function. The value $p_t(x, E)$ expresses the transition probability that a physical particle starting at position x will be found in the set E at time t. Equation (1.1) expresses the idea that a transition from the position x to the set E in time $t + s$ is composed of a transition from x to some position y in time t, followed by a transition from y to the set E in the remaining time s; the latter transition has probability $p_s(y, E)$ which depends only on y. Thus a particle "starts afresh"; this property is called the *Markov property*.

We add a point ∂ to K as the point at infinity if K is not compact, and as an isolated point if K is compact; so the space $K_\partial = K \cup \{\partial\}$ is compact.

Let $C(K)$ be the space of real-valued, bounded continuous functions on K. The space $C(K)$ is a Banach space with the supremum norm

$$\|f\| = \sup_{x \in K} |f(x)|.$$

We say that a function $f \in C(K)$ converges to zero as $x \to \partial$ if, for each $\varepsilon > 0$, there exists a compact subset E of K such that

$$|f(x)| < \varepsilon, \quad x \in K \backslash E,$$

and write $\lim_{x \to \partial} f(x) = 0$. We let

$$C_0(K) = \left\{ f \in C(K); \; \lim_{x \to \partial} f(x) = 0 \right\}.$$

The space $C_0(K)$ is a closed subspace of $C(K)$; hence it is a Banach space. Note that $C_0(K)$ may be identified with $C(K)$ if K is compact.

Now we introduce a useful convention:

Any real-valued function f on K is extended to the space $K_\partial = K \cup \{\partial\}$ by setting $f(\partial) = 0$.

From this point of view, the space $C_0(K)$ is identified with the subspace of $C(K_\partial)$ which consists of all functions f satisfying $f(\partial) = 0$, that is,

$$C_0(K) = \{ f \in C(K_\partial); \; f(\partial) = 0 \}.$$

Further we can extend a Markov transition function p_t on K to a Markov transition function p'_t on K_∂ as follows:

$$\begin{cases} p'_t(x, E) = p_t(x, E), \; x \in K, \; E \in \mathcal{B}; \\ p'_t(x, \{\partial\}) = 1 - p_t(x, K), \; x \in K; \\ p'_t(\partial, K) = 0, \; p'_t(\partial, \{\partial\}) = 1. \end{cases}$$

Intuitively, this means that a Markovian particle moves in the space K until it "dies" at which time it reaches the point ∂; hence the point ∂ is called the *terminal point*.

Now we introduce some conditions on the measures $p_t(x, \cdot)$ related to continuity in $x \in K$, for fixed $t \geq 0$.

A Markov transition function p_t is called a *Feller function* if the function

$$T_t f(x) = \int_K p_t(x, dy) f(y)$$

is a continuous function of $x \in K$ whenever f is in $C(K)$, or equivalently, if we have

$$f \in C(K) \implies T_t f \in C(K).$$

In other words, the Feller property is equivalent to saying that the measures $p_t(x, \cdot)$ depend continuously on $x \in K$ in the usual weak topology, for every fixed $t \geq 0$.

We say that p_t is a C_0-*function* if the space $C_0(K)$ is an invariant subspace of $C(K)$ for the operators T_t:

$$f \in C_0(K) \implies T_t f \in C_0(K).$$

The Feller or C_0-property deals with continuity of a Markov transition function $p_t(x, E)$ in x, and does not, by itself, have no concern with continuity in t. We give a necessary and sufficient condition on $p_t(x, E)$ in order that its associated operators $\{T_t\}_{t\geq0}$ be strongly continuous in t on the space $C_0(K)$:

$$\lim_{s\downarrow0} \|T_{t+s}f - T_tf\| = 0, \ \ f \in C_0(K).$$

A Markov transition function p_t on K is said to be *uniformly stochastically continuous* on K if the following condition is satisfied: For each $\varepsilon > 0$ and each compact $E \subset K$, we have

$$\lim_{t\downarrow0} \sup_{x\in E} [1 - p_t(x, U_\varepsilon(x))] = 0,$$

where $U_\varepsilon(x) = \{y \in K; \rho(x, y) < \varepsilon\}$ is an ε-neighborhood of x.

Then we have the following (cf. [12, Theorem 9.2.3]):

Theorem 1.1. *Let p_t be a C_0-transition function on K. Then the associated operators $\{T_t\}_{t\geq0}$, defined by*

$$(1.2) \qquad T_tf(x) = \int_K p_t(x, dy)f(y), \ \ f \in C_0(K),$$

is strongly continuous in t on $C_0(K)$ if and only if p_t is uniformly stochastically continuous on K and satisfies the following condition (L):

(L) For each $s > 0$ and each compact $E \subset K$, we have

$$\lim_{x\to\partial} \sup_{0\leq t\leq s} p_t(x, E) = 0.$$

A family $\{T_t\}_{t\geq0}$ of bounded linear operators acting on $C_0(K)$ is called a *Feller semigroup* on K if it satisfies the following three conditions:

   (i) $T_{t+s} = T_t \cdot T_s$, $t,s \geq 0$; $T_0 = I =$ the identity.

   (ii) The family $\{T_t\}$ is strongly continuous in t for $t \geq 0$:

$$\lim_{s\downarrow0} \|T_{t+s}f - T_tf\| = 0, \ \ f \in C_0(K).$$

   (iii) The family $\{T_t\}$ is non-negative and contractive on $C_0(K)$:

$$f \in C_0(K), 0 \leq f \leq 1 \text{ on } K \implies 0 \leq T_tf \leq 1 \text{ on } K.$$

The next theorem gives a characterization of Feller semigroups in terms of Markov transition functions (cf. [12, Theorem 9.2.6]):

Theorem 1.2. *If p_t is a uniformly stochastically continuous C_0-transition function on K and satisfies condition (L), then its associated operators $\{T_t\}_{t\geq0}$ form a Feller semigroup on K.*

Conversely, if $\{T_t\}_{t\geq0}$ is a Feller semigroup on K, then there exists a uniformly stochastically continuous C_0-transition p_t on K, satisfying condition (L), such that formula (1.2) holds.

1.2 Generation Theorems of Feller Semigroups

If $\{T_t\}_{t\geq 0}$ is a Feller semigroup on K, we define its *infinitesimal generator* $\mathfrak{A}$ by the formula

$$(1.3) \qquad \mathfrak{A}u = \lim_{t\downarrow 0} \frac{T_t u - u}{t} \; ,$$

provided that the limit (1.3) exists in the space $C_0(K)$. More precisely, the generator $\mathfrak{A}$ is a linear operator from the space $C_0(K)$ into itself defined as follows.

(1) The domain $D(\mathfrak{A})$ of $\mathfrak{A}$ is the set

$$D(\mathfrak{A}) = \{u \in C_0(K); \text{ the limit } (1.3) \text{ exists}\}.$$

(2) $\mathfrak{A}u = \lim_{t\downarrow 0} \frac{T_t u - u}{t}$, $u \in D(\mathfrak{A})$.

The next theorem is a version of the Hille–Yosida theorem adapted to the present context (cf. [12, Theorem 9.3.1 and Corollary 9.3.2]):

Theorem 1.3. *(i) Let $\{T_t\}_{t\geq 0}$ be a Feller semigroup on K and $\mathfrak{A}$ its infinitesimal generator. Then we have the following:*

(a) The domain $D(\mathfrak{A})$ is everywhere dense in the space $C_0(K)$.

(b) For each $\alpha > 0$, the equation $(\alpha I - \mathfrak{A})u = f$ has a unique solution u in $D(\mathfrak{A})$ for any $f \in C_0(K)$. Hence, for each $\alpha > 0$, the Green operator $(\alpha I - \mathfrak{A})^{-1} : C_0(K) \longrightarrow C_0(K)$ can be defined by the formula

$$u = (\alpha I - \mathfrak{A})^{-1}f, \;\; f \in C_0(K).$$

(c) For each $\alpha > 0$, the operator $(\alpha I - \mathfrak{A})^{-1}$ is non-negative on the space $C_0(K)$:

$$f \in C_0(K), \, f \geq 0 \text{ on } K \implies (\alpha I - \mathfrak{A})^{-1}f \geq 0 \text{ on } K.$$

(d) For each $\alpha > 0$, the operator $(\alpha I - \mathfrak{A})^{-1}$ is bounded on the space $C_0(K)$ with norm

$$\|(\alpha I - \mathfrak{A})^{-1}\| \leq \frac{1}{\alpha} \; .$$

(ii) Conversely, if $\mathfrak{A}$ is a linear operator from the space $C_0(K)$ into itself satisfying condition (a) and if there is a constant $\alpha_0 \geq 0$ such that, for all $\alpha > \alpha_0$, conditions (b) through (d) are satisfied, then $\mathfrak{A}$ is the infinitesimal generator of some Feller semigroup $\{T_t\}_{t\geq 0}$ on K.

Corollary 1.4. *Let K be a compact metric space and let $\mathfrak{A}$ be the infinitesimal generator of a Feller semigroup on K. Assume that the constant function 1 belongs to the domain $D(\mathfrak{A})$ of $\mathfrak{A}$ and that we have for some constant c*

$$\mathfrak{A}1 \leq -c \text{ on } K.$$

Then the operator $\mathfrak{A}' = \mathfrak{A} + cI$ is the infinitesimal generator of some Feller semigroup on K.

Although Theorem 1.3 tells us precisely when a linear operator $\mathfrak{A}$ is the infinitesimal generator of some Feller semigroup, it is usually difficult to verify conditions (b) through (d). So we give useful criteria in terms of the *maximum principle* (cf. [12, Theorem 9.3.3 and Corollary 9.3.4]):

Theorem 1.5. *Let K be a compact metric space. Then we have the following assertions:*

(i) Let B be a linear operator from the space $C(K) = C_0(K)$ into itself, and assume that:

(α) The domain $D(B)$ of B is everywhere dense in the space $C(K)$.

(β) There exists an open and dense subset K_0 of K such that if $u \in D(B)$ takes a positive maximum at a point x_0 of K_0, then we have

$$Bu(x_0) \leq 0.$$

Then the operator B is closable in the space $C(K)$.

(ii) Let B be as in part (i), and further assume that:

(β') If $u \in D(B)$ takes a positive maximum at a point x' of K, then we have

$$Bu(x') \leq 0.$$

(γ) For some $\alpha_0 \geq 0$, the range $R(\alpha_0 I - B)$ of $\alpha_0 I - B$ is everywhere dense in the space $C(K)$.

Then the minimal closed extension $\bar{B}$ of B is the infinitesimal generator of some Feller semigroup on K.

Corollary 1.6. *Let $\mathfrak{A}$ be the infinitesimal generator of a Feller semigroup on a compact metric space K and M a bounded linear operator on the space $C(K)$ into itself. Assume that either M or $\mathfrak{A}' = \mathfrak{A} + M$ satisfies condition (β'). Then the operator $\mathfrak{A}'$ is the infinitesimal generator of some Feller semigroup on K.*

CHAPTER II

THEORY OF PSEUDO–DIFFERENTIAL OPERATORS

In this chapter we present a brief description of the basic concepts and results of the Hölder space theory of pseudo-differential operators which will be used in subsequent chapters. In particular, we introduce the notion of transmission property, due to Boutet de Monvel [2], which is a condition about symbols in the normal direction at the boundary. Furthermore, we prove an existence and uniqueness theorem for a class of pseudo-differential operators which enters naturally in the construction of Feller semigroups. For detailed studies of pseudo-differential operators, the reader is referred to Kumano-go [6], Rempel-Schulze [8] and Taylor [15].

2.1 Function Spaces

Let Ω be an open subset of Euclidean space $\mathbf{R}^n$. A Lebesgue measurable function u on Ω is said to be *essentially bounded* if there exists a constant $C > 0$ such that $|u(x)| \leq C$ almost everywhere (a.e.) in Ω. We define

$$\text{ess sup}_{x\in\Omega} |u(x)| = \inf\{C; |u(x)| \leq C \text{ a.e. in } \Omega\}.$$

We let

$$L^\infty(\Omega) = \text{the space of equivalence classes of essentially bounded,}$$
$$\text{Lebesgue measurable functions on } \Omega.$$

The space $L^\infty(\Omega)$ is a Banach space with the norm

$$\|u\|_\infty = \text{ess sup}_{x\in\Omega} |u(x)|.$$

If m is a non-negative integer, we let

$$W^{m,\infty}(\Omega) = \text{the space of equivalence classes of functions}$$
$$u \in L^\infty(\Omega) \text{ all of whose derivatives } \partial^\alpha u,$$
$$|\alpha| \leq m, \text{ in the sense of distributions are}$$
$$\text{in } L^\infty(\Omega).$$

The space $W^{m,\infty}(\Omega)$ is a Banach space with the norm

$$\|u\|_{m,\infty} = \sum_{|\alpha|\leq m} \|\partial^\alpha u\|_\infty.$$

Here and in the following we use the shorthand

$$\partial_j = \frac{\partial}{\partial x_j}, \; 1 \leq j \leq n,$$

$$\partial^\alpha = \partial_1^{\alpha_1} \ldots \partial_n^{\alpha_n} , \ \alpha = (\alpha_1, \ldots, \alpha_n),$$

for derivatives on $\mathbf{R}^n$.

We remark that

$$W^{0,\infty}(\Omega) = L^\infty(\Omega) \ ; \ \|\cdot\|_{0,\infty} = \|\cdot\|_\infty.$$

Now we let

$C(\Omega) = $ the space of continuous functions on Ω.

If k is a positive integer, we let

$C^k(\Omega) = $ the space of functions of class C^k on Ω.

Further we let

$C(\bar\Omega) = $ the space of functions in $C(\Omega)$ having continuous extensions
to the closure $\bar\Omega$ of Ω.

If k is a positive integer, we let

$C^k(\bar\Omega) = $ the space of functions in $C^k(\Omega)$ all of whose derivatives
of order $\leq k$ have continuous extensions to $\bar\Omega$.

The space $C^k(\bar\Omega)$ is a Banach space with the norm

$$\|u\|_{C^k(\bar\Omega)} = \sup_{\substack{x \in \bar\Omega \\ |\alpha| \leq k}} |\partial^\alpha u(x)|.$$

Let $0 < \theta < 1$. A function u defined on Ω is said to be *Hölder continuous* with exponent θ if the quantity

$$[u]_{\theta;\Omega} = \sup_{\substack{x,y \in \Omega \\ x \neq y}} \frac{|u(x) - u(y)|}{|x - y|^\theta}$$

is finite. We say that u is *locally Hölder continuous* with exponent θ if it is Hölder continuous with exponent θ on compact subsets of Ω.

We let

$C^\theta(\Omega) = $ the space of functions in $C(\Omega)$ which are locally Hölder
continuous with exponent θ on Ω.

If k is a positive integer, we let

$C^{k+\theta}(\Omega) = $ the space of functions in $C^k(\Omega)$ all of whose k-th order
derivatives are locally Hölder continuous with exponent θ

on Ω.

Further we let

$$C^\theta(\bar\Omega) = \text{the space of functions in } C(\bar\Omega) \text{ which are Hölder}$$
$$\text{continuous with exponent } \theta \text{ on } \bar\Omega.$$

If k is a positive integer, we let

$$C^{k+\theta}(\bar\Omega) = \text{the space of functions in } C^k(\bar\Omega) \text{ all of whose } k\text{-th order}$$
$$\text{derivatives are Hölder continuous with exponent } \theta$$
$$\text{on } \bar\Omega.$$

The space $C^{k+\theta}(\bar\Omega)$ is a Banach space with the norm

$$\|u\|_{C^{k+\theta}(\bar\Omega)} = \|u\|_{C^k(\bar\Omega)} + \sup_{|\alpha|=k} [\partial^\alpha u]_{\theta;\bar\Omega}.$$

If M is an n-dimensional compact C^∞ manifold without boundary and m is a non-negative integer, then the spaces $W^{m,\infty}(M)$ and $C^{m+\theta}(M)$ are defined respectively to be locally the spaces $W^{m,\infty}(\mathbf{R}^n)$ and $C^{m+\theta}(\mathbf{R}^n)$, upon using local coordinate systems flattening out M, together with a partition of unity. The norms of the spaces $W^{m,\infty}(M)$ and $C^{m+\theta}(M)$ will be denoted by $\|\cdot\|_{m,\infty}$ and $\|\cdot\|_{C^{m+\theta}(M)}$, respectively.

We recall the following results (cf. Triebel [16]):

I) If k is a positive integer, then we have

$$W^{k,\infty}(M) = \{\varphi \in C^{k-1}(M); \max_{|\alpha|\le k-1} \sup_{\substack{x,y\in M \\ x\ne y}} \frac{|\partial^\alpha\varphi(x) - \partial^\alpha\varphi(y)|}{|x-y|} < \infty\},$$

where $|x - y|$ is the geodesic distance between x and y with respect to the Riemannian metric of M.

II) The space $C^{k+\theta}(M)$ is a *real interpolation space* between the spaces $W^{k,\infty}(M)$ and $W^{k+1,\infty}(M)$; more precisely we have

$$C^{k+\theta}(M) = \left(W^{k,\infty}(M), W^{k+1,\infty}(M)\right)_{\theta,\infty}$$
$$= \left\{u \in W^{k,\infty}(M); \sup_{t>0} \frac{K(t,u)}{t^\theta} < \infty\right\},$$

where

$$K(t,u) = \inf_{u=u_0+u_1} \left(\|u_0\|_{k,\infty} + t\|u_1\|_{k+1,\infty}\right).$$

2.2 Pseudo-Differential Operators

Let Ω be an open subset of Euclidean space $\mathbf{R}^n$. If $m \in \mathbf{R}$ and $0 \leq \delta < \rho \leq 1$, we let

$$
\begin{aligned}
S^m_{\rho,\delta}(\Omega \times \mathbf{R}^N) = &\text{ the set of all functions } a \in C^\infty(\Omega \times \mathbf{R}^N) \text{ with the property} \\
&\text{that, for any compact } K \subset \Omega \text{ and multi-indices } \alpha,\ \beta,\ \text{there} \\
&\text{exists a constant } C_{K,\alpha,\beta} > 0 \text{ such that we have for all} \\
&x \in K \text{ and } \theta \in \mathbf{R}^N \\
&\left| \partial_\theta^\alpha \partial_x^\beta a(x,\theta) \right| \leq C_{K,\alpha,\beta}(1 + |\theta|)^{m-\rho|\alpha|+\delta|\beta|}.
\end{aligned}
$$

The elements of $S^m_{\rho,\delta}(\Omega \times \mathbf{R}^N)$ are called *symbols* of order m.

We set

$$
S^{-\infty}(\Omega \times \mathbf{R}^N) = \cap_{m \in \mathbf{R}} S^m_{\rho,\delta}(\Omega \times \mathbf{R}^N).
$$

If $a_j \in S^{m_j}_{\rho,\delta}(\Omega \times \mathbf{R}^N)$ is a sequence of symbols of decreasing order, then there exists a symbol $a \in S^{m_0}_{\rho,\delta}(\Omega \times \mathbf{R}^N)$, unique modulo $S^{-\infty}(\Omega \times \mathbf{R}^N)$, such that we have for all $k > 0$:

$$
a - \sum_{j=0}^{k-1} a_j \in S^{m_k}_{\rho,\delta}(\Omega \times \mathbf{R}^N).
$$

In this case, we write

$$
a \sim \sum_{j=0}^{\infty} a_j \ .
$$

The formal sum $\sum_j a_j$ is called an asymptotic expansion of a.

A symbol $a(x,\theta) \in S^m_{1,0}(\Omega \times \mathbf{R}^N)$ is said to be *classical* if there exist C^∞ functions $a_j(x,\theta)$, positively homogeneous of degree $m - j$ in θ for $|\theta| \geq 1$, such that

$$
a \sim \sum_{j=0}^{\infty} a_j \ .
$$

The homogeneous function a_0 of degree m is called the principal part of a.

We let

$$
S^m_{cl}(\Omega \times \mathbf{R}^N) = \text{ the set of all classical symbols of order } m.
$$

Let Ω be an open subset of $\mathbf{R}^n$ and $m \in \mathbf{R}$. A *pseudo-differential operator* of order m on Ω is a Fourier integral operator of the form

$$
Au(x) = \iint_{\Omega \times \mathbf{R}^n} e^{i(x-y)\cdot\xi} a(x,y,\xi) u(y)\, dy d\xi, \ \ u \in C_0^\infty(\Omega),
$$

with some $a \in S^m_{\rho,\delta}(\Omega \times \Omega \times \mathbf{R}^n)$. Here the integral is taken in the sense of *oscillatory integrals*.

We let

$$L_{\rho,\delta}^m(\Omega) = \text{the set of all pseudo-differential operators of order } m \text{ on } \Omega,$$

and set

$$L^{-\infty}(\Omega) = \cap_{m \in \mathbf{R}} L_{\rho,\delta}^m(\Omega).$$

Recall that a continuous linear operator $A : C_0^\infty(\Omega) \longrightarrow \mathcal{D}'(\Omega)$ is said to be *properly supported* if the following two conditions are satisfied:

(a) For any compact subset K of Ω, there exists a compact subset K' of Ω such that

$$\text{supp } v \subset K \implies \text{supp } Av \subset K'.$$

(b) For any compact subset K' of Ω, there exists a compact subset K of Ω such that

$$\text{supp } v \cap K = \emptyset \implies \text{supp } Av \cap K' = \emptyset.$$

If $A \in L_{\rho,\delta}^m(\Omega)$, one can choose a properly supported operator $A_0 \in L_{\rho,\delta}^m(\Omega)$ such that $A - A_0 \in L^{-\infty}(\Omega)$, and define

$$\sigma(A) = \text{the equivalence class of the complete symbol of } A_0$$

$$\text{in the factor class } S_{\rho,\delta}^m(\Omega \times \mathbf{R}^n)/S^{-\infty}(\Omega \times \mathbf{R}^n).$$

The equivalence class $\sigma(A)$ does not depend on the operator A_0 chosen, and is called the *complete symbol* of A.

We shall often identify the complete symbol $\sigma(A)$ with a representative in the class $S_{\rho,\delta}^m(\Omega \times \mathbf{R}^n)$ for notational convenience, and call any member of $\sigma(A)$ a complete symbol of A.

A pseudo-differential operator $A \in L_{1,0}^m(\Omega)$ is said to be *classical* if its complete symbol $\sigma(A)$ has a representative in the class $S_{cl}^m(\Omega \times \mathbf{R}^n)$.

We let

$$L_{cl}^m(\Omega) = \text{the set of all classical pseudo-differential operators of order } m$$
$$\text{on } \Omega.$$

If $A \in L_{cl}^m(\Omega)$, then the principal part of $\sigma(A)$ has a canonical representative $\sigma_A(x, \xi) \in C^\infty(\Omega \times (\mathbf{R}^n \backslash \{0\}))$ which is positively homogeneous of degree m in the variable ξ. The function $\sigma_A(x, \xi)$ is called the *homogeneous principal symbol* of A.

Now we define the concept of a pseudo-differential operator on a manifold, and transfer all the machinery of pseudo-differential operators to manifolds.

Let M be an n-dimensional, *compact* C^∞ manifold without boundary and $1 - \rho \leq \delta < \rho \leq 1$. A continuous linear operator $A : C^\infty(M) \longrightarrow C^\infty(M)$ is called a *pseudo-differential operator* of order $m \in \mathbf{R}$ if it satisfies the following two conditions:

(i) The distribution kernel of A is of class C^∞ off the diagonal $\Delta_M = \{(x, x) ; x \in M\}$ in $M \times M$.

(ii) For any chart (U, χ) on M, the mapping

$$A_\chi : C_0^\infty(\chi(U)) \longrightarrow C^\infty(\chi(U))$$
$$u \longmapsto A(u \circ \chi) \circ \chi^{-1}$$

belongs to the class $L_{\rho,\delta}^m(\chi(U))$.

We let

$$L_{\rho,\delta}^m(M) = \text{the set of all pseudo-differential operators of order } m \text{ on } M,$$

and set

$$L^{-\infty}(M) = \cap_{m\in\mathbf{R}} L_{\rho,\delta}^m(M).$$

Some results about pseudo-differential operators on $\mathbf{R}^n$ are also true for pseudo-differential operators on M, since pseudo-differential operators on M are defined to be locally pseudo-differential operators on $\mathbf{R}^n$.

For example, we have the following results:

1) A pseudo-differential operator A extends to a continuous linear operator $A : \mathcal{D}'(M) \longrightarrow \mathcal{D}'(M)$.

2) sing supp $Au \subset$ sing supp u, $u \in \mathcal{D}'(M)$.

3) A continuous linear operator $A : C^\infty(M) \longrightarrow \mathcal{D}'(M)$ is a regularizer if and only if it is in $L^{-\infty}(M)$.

4) The class $L_{\rho,\delta}^m(M)$, $1 - \rho \leq \delta < \rho \leq 1$, is stable under the operations of composition of operators and taking the transpose or adjoint of an operator.

5) A pseudo-differential operator $A \in L_{1,0}^m(M)$ extends to a continuous linear operator $A : C^{k+\theta}(M) \longrightarrow C^{k-m+\theta}(M)$ for any integer $k \geq m$.

A pseudo-differential operator $A \in L_{1,0}^m(M)$ is said to be *classical* if, for any chart (U,χ) on M, the mapping $A_\chi : C_0^\infty(\chi(U)) \longrightarrow C^\infty(\chi(U))$ belongs to the class $L_{cl}^m(\chi(U))$.

We let

$$L_{cl}^m(M) = \text{the set of all classical pseudo-differential operators of order}$$
$$m \text{ on } M.$$

We observe that

$$L^{-\infty}(M) = \cap_{m\in\mathbf{R}} L_{cl}^m(M).$$

Let $A \in L_{cl}^m(M)$. If (U,χ) is a chart on M, there is associated a homogeneous principal symbol $\sigma_{A_\chi} \in C^\infty(\chi(U) \times (\mathbf{R}^n\backslash\{0\}))$. Then, by smoothly patching together the functions σ_{A_χ}, one can obtain a C^∞ function $\sigma_A(x,\xi)$ on $T^*(M)\backslash\{0\} = \{(x,\xi) \in T^*(M); \xi \neq 0\}$, which is positively homogeneous of degree m in the variable ξ. The function $\sigma_A(x,\xi)$ is called the *homogeneous principal symbol* of A.

A classical pseudo-differential operator $A \in L_{cl}^m(M)$ is said to be *elliptic* of order m if its homogeneous principal symbol $\sigma_A(x,\xi)$ does not vanish on the bundle $T^*(M)\backslash\{0\}$ of non-zero cotangent vectors.

Then we have the following:

6) An operator $A \in L_{cl}^m(M)$ is elliptic if and only if there exists a parametrix $B \in L_{cl}^{-m}(M)$ for A:

$$\begin{cases} AB \equiv I \mod L^{-\infty}(M), \\ BA \equiv I \mod L^{-\infty}(M). \end{cases}$$

Finally we introduce the notion of transmission property, due to Boutet de Monvel [2], which is a condition about symbols in the normal direction at the boundary.

We let

$$S_{1,0}^m(\overline{\mathbf{R}_+^n} \times \overline{\mathbf{R}_+^n} \times \mathbf{R}^n) = \text{the space of symbols in } S_{1,0}^m(\mathbf{R}_+^n \times \mathbf{R}_+^n \times \mathbf{R}^n)$$
$$\text{which have an extension in } S_{1,0}^m(\mathbf{R}^n \times \mathbf{R}^n \times \mathbf{R}^n).$$

We say that a symbol $a(x, y, \xi) \in S_{1,0}^m(\overline{\mathbf{R}_+^n} \times \overline{\mathbf{R}_+^n} \times \mathbf{R}^n)$ has the *transmission property* with respect to the boundary $\mathbf{R}^{n-1}$ if the function $a(x, y, \xi)$ and all its derivatives admit an expansion of the form:

$$\left(\frac{\partial}{\partial x}\right)^\alpha \left(\frac{\partial}{\partial y}\right)^\beta a(x, y, \xi) \Bigg|_{\substack{x=y \\ x_n=0}}$$
$$= \sum_{j=0}^m b_j(x', \xi') \xi_n^j + \sum_{k=-\infty}^\infty a_k(x', \xi') \frac{(<\xi'> -i\xi_n)^k}{(<\xi'> +i\xi_n)^{k+1}},$$

where $b_j \in S_{1,0}^{m-j}(\mathbf{R}^{n-1} \times \mathbf{R}^{n-1})$ and the a_k form a rapidly decreasing sequence in $S_{1,0}^{m+1}(\mathbf{R}^{n-1} \times \mathbf{R}^{n-1})$ with respect to k, and $<\xi'> = (1 + |\xi'|^2)^{1/2}$.

We let

$$L_{1,0}^m(\overline{\mathbf{R}_+^n}) = \text{the space of pseudo-differential operators in } L_{1,0}^m(\mathbf{R}_+^n) \text{ which can}$$
$$\text{be extended to a pseudo-differential operator in } L_{1,0}^m(\mathbf{R}^n).$$

A pseudo-differential operator $A \in L_{1,0}^m(\overline{\mathbf{R}_+^n})$ is said to have the *transmission property* with respect to the boundary $\mathbf{R}^{n-1}$ if any complete symbol of A has the transmission property with respect to the boundary $\mathbf{R}^{n-1}$.

If A is a pseudo-differential operator in $L_{1,0}^m(\overline{\mathbf{R}_+^n})$, then we define a new operator

$$A_{\mathbf{R}_+^n} : C_0^\infty\left(\overline{\mathbf{R}_+^n}\right) \longrightarrow C^\infty\left(\mathbf{R}_+^n\right)$$
$$u \longmapsto (A\tilde{u})|_{\mathbf{R}_+^n},$$

where $\tilde{u}$ is the extension of u to $\mathbf{R}^n$ by 0 outside $\mathbf{R}_+^n$.

The transmission property implies that if u is of class C^∞ up to the boundary, then so is $A_{\mathbf{R}_+^n} u$. More precisely, we have the following results:

I) If a pseudo-differential operator $A \in L_{1,0}^m(\overline{\mathbf{R}_+^n})$ has the transmission property with respect to the boundary $\mathbf{R}^{n-1}$, then $A_{\mathbf{R}_+^n}$ maps $C_0^\infty\left(\overline{\mathbf{R}_+^n}\right)$ continuously into $C^\infty\left(\overline{\mathbf{R}_+^n}\right)$.

II) If a pseudo-differential operator $A \in L_{1,0}^m(\overline{\mathbf{R}_+^n})$ has the transmission property, then $A_{\mathbf{R}_+^n}$ maps $C_{comp}^{k+\theta}\left(\overline{\mathbf{R}_+^n}\right)$ continuously into $C_{loc}^{k-m+\theta}\left(\overline{\mathbf{R}_+^n}\right)$ for any integer $k \geq m$. Here $C_{comp}^{k+\theta}\left(\overline{\mathbf{R}_+^n}\right)$ is the space of Hölder continuous functions with compact support and $C_{loc}^{k-m+\theta}\left(\overline{\mathbf{R}_+^n}\right)$ is the space of locally Hölder continuous functions, respectively.

We remark that the notion of transmission property is invariant under a change of coordinates which preserves the boundary. Hence this notion can be transferred to manifolds with boundary as follows.

Let Ω be an n-dimensional, *compact* C^∞ manifold with boundary Γ, and let M be an n-dimensional, paracompact C^∞ manifold without boundary such that Ω is a relatively compact open subset of M. The notion of transmission property can be extended to the class $L_{1,0}^m(M)$, upon using local coordinate systems flattening out the boundary Γ.

Then we have the following results:

I') If a pseudo-differential operator $A \in L_{1,0}^m(M)$ has the transmission property with respect to the boundary Γ, then the operator

$$A_\Omega : C^\infty(\bar{\Omega}) \longrightarrow C^\infty(\Omega)$$
$$u \longmapsto (A\tilde{u})|_\Omega$$

maps $C^\infty(\bar{\Omega})$ continuously into $C^\infty(\bar{\Omega})$. Here $\bar{\Omega} = \Omega \cup \Gamma$ and $\tilde{u}$ is the extension of u to M by 0 outside Ω.

II') If a pseudo-differential operator $A \in L_{1,0}^m(M)$ has the transmission property, then the operator A_Ω maps $C^{k+\theta}(\bar{\Omega})$ continuously into $C^{k-m+\theta}(\bar{\Omega})$ for any integer $k \geq m$.

2.3 Unique Solvability Theorem for Pseudo-Differential Operators

The next result will play an essential role in the construction of Feller semigroups in Chapters III and IV.

Theorem 2.1. *Let T be a classical pseudo-differential operator of second order on an n-dimensional compact C^∞ manifold M without boundary such that*

$$T = P + S,$$

where:

a) *The operator P is a second-order degenerate elliptic differential operator on M with non-positive principal symbol, and $P1 \leq 0$ on M.*

b) *The operator S is a classical pseudo-differential operator of order $2 - \kappa$, $\kappa > 0$, on M and its distribution kernel $s(x, y)$ is non-negative off the diagonal $\Delta_M = \{(x, x)\,; x \in M\}$ in $M \times M$.*

c) *$T1 = P1 + S1 \leq 0$ on M.*

Then, for each integer $k \geq 1$, there exists a constant $\lambda = \lambda(k) > 0$ such that for any $f \in C^{k+\theta}(M)$ one can find a function $\varphi \in C^{k+\theta}(M)$ satisfying

$$(T - \lambda I)\varphi = f \text{ on } M,$$

and

$$\|\varphi\|_{C^{k+\theta}(M)} \leq C_{k+\theta}(\lambda)\|f\|_{C^{k+\theta}(M)}\,.$$

Here $C_{k+\theta}(\lambda) > 0$ is a constant independent of f.

Proof. We prove Theorem 2.1 by using a method of *elliptic regularization* (cf. Oleĭnik-Radkevič [7, Chapter I]), just as in the proof of Théorème 4.5 of Cancelier [3]. So we only give a sketch of the proof.

1) We recall the following results:

Theorem 2.2. Let $T = P + S$ be a classical pseudo-differential operator of second order on M as in Theorem 2.1. Assume that

$$T1 = P1 + S1 < 0 \text{ on } M.$$

Then we have for all $\varphi \in C^2(M)$

$$\|\varphi\|_{C(M)} \le \left(\frac{1}{\min_M(-T1)} \right) \|T\varphi\|_{C(M)}.$$

Theorem 2.2 is a compact manifold version of Theorem A.2 in Appendix.

Theorem 2.3. Let $T = P + S$ be a classical pseudo-differential operator of second order on M as in Theorem 2.1. Assume that the operator T is elliptic on M and satisfies the condition

$$T1 = P1 + S1 < 0 \text{ on } M.$$

Then, for each integer $k \ge 0$, the operator

$$T : C^{k+2+\theta}(M) \longrightarrow C^{k+\theta}(M)$$

is bijective.

Since T is elliptic and its principal symbol is *real*, it follows from an application of Corollary 6.7.12 of Taira [12] that

$$\text{ind } T = \dim N(T) - \text{codim } R(T) = 0.$$

But Theorem 2.2 tells us that T is injective, that is, $\dim N(T) = 0$. Hence we obtain that $\text{codim } R(T) = 0$, which proves that T is surjective.

2) First we prove Theorem 2.1 for the space $W^{1,\infty}(M)$:

Claim I. There exists a constant $\lambda = \lambda(1) > 0$ such that for any $f \in W^{1,\infty}(M)$ one can find a function $\varphi \in W^{1,\infty}(M)$ satisfying

$$(T - \lambda I)\varphi = f \text{ on } M,$$

and

$$\|\varphi\|_{1,\infty} \le C_1(\lambda)\|f\|_{1,\infty}.$$

Here $C_1(\lambda) > 0$ is a constant independent of f.

Proof. 2-i) Let $\{(U_\alpha, \chi_\alpha)\}_{\alpha=1}^N$ be a finite open covering of M by local charts, and let $\{\sigma_\alpha\}_{\alpha=1}^N$ be a family of non-negative functions in $C^\infty(M \times M)$ such that

$$\text{supp } \sigma_\alpha \subset U_\alpha \times U_\alpha,$$

and

$$\sum_{\alpha=1}^N \sigma_\alpha(x, y) = 1 \text{ in a neighborhood of the diagonal } \Delta_M = \{(x, x); x \in M\}.$$

Then it is easy to see that the operator $T = P+S$ can be written in local coordinates $(x_1, \ldots, x_n)$ in the following form:

$$T\varphi(x) = \sum_{i,j=1}^{n} \alpha^{ij}(x)\frac{\partial^2\varphi}{\partial x_i \partial x_j}(x) + \sum_{i=1}^{n}\beta^i(x)\frac{\partial\varphi}{\partial x_i}(x) + \gamma(x)\varphi(x)$$

$$+ \int_M s(x,y)\left[\varphi(y) - \sigma(x,y)\left(\varphi(x) + \sum_{i=1}^{n}(y_i - x_i)\frac{\partial\varphi}{\partial x_i}(x)\right)\right]dy.$$

Here:

a) The operator $\sum_{i,j=1}^{n}\alpha^{ij}\partial^2/\partial x_i \partial x_j$ is the principal part of P; more precisely, the α^{ij} are the components of a C^∞ symmetric contravariant tensor of type $\binom{2}{0}$ on M satisfying

$$\sum_{i,j=1}^{n}\alpha^{ij}(x)\xi_i\xi_j \geq 0, \quad x \in M, \ \xi = \sum_{j=1}^{n}\xi_j\,dx_j \in T_x^*(M),$$

where $T_x^*(M)$ is the cotangent space of M at x.

b) $\sigma(x,y) = \sum_{\alpha=1}^{N}\sigma_\alpha(x,y)$.

c) The density dy is a strictly positive density on M.

d) $T1(x) = \gamma(x) + \int_M s(x,y)[1 - \sigma(x,y)]dy \leq 0$ on M.

Furthermore, we remark that there exists a constant $C > 0$ such that the distribution kernel $s(x,y)$ of $S \in L_{cl}^{2-\kappa}(M)$, $\kappa > 0$, satisfies the estimate

$$|s(x,y)| \leq \frac{C}{|x-y|^{n+2-\kappa}}, \quad (x,y) \in (M \times M)\backslash\Delta_M,$$

where $|x-y|$ is the geodesic distance between x and y with respect to the Riemannian metric of M (cf. Coifman-Meyer [4, Chapitre IV, Proposition 1]). Hence we find that the integral

$$\int_M s(x,y)\left[\varphi(y) - \sigma(x,y)\left(\varphi(x) + \sum_{i=1}^{n}(y_i - x_i)\frac{\partial\varphi}{\partial x_i}(x)\right)\right]dy$$

is absolutely convergent, since $\kappa > 0$ and $\sigma(x,y) = 1$ in a neighborhood of the diagonal Δ_M.

Now, if $\varphi \in C^1(M)$, we define a continuous function $B_T(\varphi, \varphi)$ on M by the formula

$$B_T(\varphi, \varphi)(x) = 2\sum_{i,j=1}^{n}\alpha^{ij}(x)\frac{\partial\varphi}{\partial x_i}(x)\frac{\partial\varphi}{\partial x_j}(x)$$

$$+ \int_M s(x,y)\left(\varphi(y) - \varphi(x)\right)^2 dy - T1(x)\cdot\varphi(x)^2, \quad x \in M.$$

We remark that the function $B(\varphi, \varphi)$ is *non-negative* on M for all $\varphi \in C^1(M)$.

The next result may be proved just as in the proof of Théorème 4.1 of Cancelier [3].

Lemma 2.4. *Let $\{X_j\}_{j=1}^r$ be a family of real C^∞ vector fields on M such that the X_j span the tangent space $T_x(M)$ at each point x of M. If $\varphi \in C^\infty(M)$, we let*

$$p_1(x) = \sum_{j=1}^r |X_j\varphi(x)|^2, \ x \in M,$$

and

$$R_1(x) = Tp_1(x) - \sum_{j=1}^r B_T(X_j\varphi, X_j\varphi)(x), \ x \in M.$$

Then, for each $\eta > 0$, there exist constants $\beta_0 > 0$ and $\beta_1 > 0$ such that we have for all $\varphi \in C^\infty(M)$

$$(2.1) \qquad |R_1(x)| \leq \eta \sum_{j=1}^r B_T(X_j\varphi, X_j\varphi)(x) + \beta_0\|\varphi\|_{C(M)}^2$$

$$+ \beta_1\|\varphi\|_{C^1(M)}^2 + \frac{1}{2}\|T\varphi\|_{C^1(M)}^2, \ x \in M.$$

Remark 2.5. The constants β_0 and β_1 are *uniform* for the operators $T + \varepsilon\Lambda - \lambda I$, $0 \leq \varepsilon \leq 1$, $\lambda \geq 0$, where Λ is a second-order *elliptic* differential operator on M defined by the formula:

$$\Lambda = -\sum_{j=1}^r X_j^* X_j = \sum_{j=1}^r X_j^2 + \sum_{j=1}^r \operatorname{div} X_j \cdot X_j.$$

2-ii) First let f be an arbitrary element of $C^\infty(M)$. Since the operator $T + \varepsilon\Lambda - \lambda I$ is elliptic for all $\varepsilon > 0$ and $(T + \varepsilon\Lambda - \lambda I)1 = T1 - \lambda \leq -\lambda < 0$ on M for $\lambda > 0$, it follows from an application of Theorem 2.3 that one can find a unique function $\varphi_\varepsilon \in C^\infty(M)$ such that

$$(T + \varepsilon\Lambda - \lambda I)\varphi_\varepsilon = f \text{ on } M.$$

Furthermore, applying Theorem 2.2 to the operator $T + \varepsilon\Lambda - \lambda I$, we obtain that

$$(2.2) \qquad \|\varphi_\varepsilon\|_{C(M)} \leq \frac{1}{\lambda}\|f\|_{C(M)},$$

since $\min_M \left(-(T + \varepsilon\Lambda - \lambda I)1\right) \geq \lambda$.

Let x_0 be a point of M at which the function

$$p_1^\varepsilon(x) = \sum_{j=1}^r |X_j\varphi_\varepsilon(x)|^2$$

attains its positive maximum. Then we have

$$\Lambda p_1^\varepsilon(x_0) = \left(\sum_{j=1}^r X_j^2\right) p_1^\varepsilon(x_0) \leq 0,$$

and also

$$
\begin{aligned}
Tp_1^\varepsilon(x_0) &= \sum_{i,j=1}^{n} \alpha^{ij}(x_0)\frac{\partial^2 p_1^\varepsilon}{\partial x_i \partial x_j}(x_0) + \gamma(x_0)p_1^\varepsilon(x_0) \\
&\quad + \int_M s(x_0,y)[p_1^\varepsilon(y) - \sigma(x_0,y)p_1^\varepsilon(x_0)]dy \\
&\leq \left(\gamma(x_0) + \int_M s(x_0,y)[1 - \sigma(x_0,y)]dy \right) p_1^\varepsilon(x_0) \\
&\quad + \int_M s(x_0,y)[p_1^\varepsilon(y) - p_1^\varepsilon(x_0)]dy \\
&\leq T1(x_0) \cdot p_1^\varepsilon(x_0).
\end{aligned}
$$

Hence, using inequality (2.1) with $\eta = 1/2$ and inequality (2.2), we obtain that

$$
\begin{aligned}
\lambda p_1^\varepsilon(x_0) &\leq (\lambda - T1(x_0))p_1^\varepsilon(x_0) - \varepsilon \Lambda p_1^\varepsilon(x_0) \\
&\leq (\lambda - T - \varepsilon \Lambda)p_1^\varepsilon(x_0) \\
&= -\left((T + \varepsilon \Lambda - \lambda)p_1^\varepsilon(x_0) - \sum_{j=1}^{r} B_{T+\varepsilon\Lambda-\lambda I}(X_j \varphi_\varepsilon, X_j \varphi_\varepsilon)(x_0) \right) \\
&\quad - \sum_{j=1}^{r} B_{T+\varepsilon\Lambda-\lambda I}(X_j \varphi_\varepsilon, X_j \varphi_\varepsilon)(x_0) \\
&\leq -\frac{1}{2}\sum_{j=1}^{r} B_{T+\varepsilon\Lambda-\lambda I}(X_j \varphi_\varepsilon, X_j \varphi_\varepsilon)(x_0) + \beta_0 \|\varphi_\varepsilon\|_{C(M)}^2 \\
&\quad + \beta_1 \|\varphi_\varepsilon\|_{C^1(M)}^2 + \frac{1}{2}\|f\|_{C^1(M)}^2 \\
&\leq \frac{\beta_0}{\lambda^2}\|f\|_{C(M)}^2 + \beta_1\|\varphi_\varepsilon\|_{C^1(M)}^2 + \frac{1}{2}\|f\|_{C^1(M)}^2.
\end{aligned}
$$

This proves that

$$
(2.3) \qquad
\begin{aligned}
(\lambda - \beta_1)\|\varphi_\varepsilon\|_{C^1(M)}^2 &\leq \lambda\left(\|\varphi_\varepsilon\|_{C(M)}^2 + p_1^\varepsilon(x_0) \right) - \beta_1\|\varphi_\varepsilon\|_{C^1(M)}^2 \\
&\leq \frac{1}{\lambda}\|f\|_{C(M)}^2 + \frac{\beta_0}{\lambda^2}\|f\|_{C(M)}^2 + \frac{1}{2}\|f\|_{C^1(M)}^2.
\end{aligned}
$$

Here we remark (cf. Remark 2.5) that the constants β_0 and β_1 are independent of $\varepsilon > 0$ and $\lambda > 0$. Therefore, if $\lambda > 0$ is so large that

$$
\lambda > \beta_1,
$$

then it follows from inequality (2.3) that

$$
(2.4) \qquad \|\varphi_\varepsilon\|_{C^1(M)}^2 \leq C(\lambda)\|f\|_{C^1(M)}^2,
$$

where $C(\lambda) > 0$ is a constant *independent* of $\varepsilon > 0$.

2-iii) Now let f be an arbitrary element of $W^{1,\infty}(M)$. Then one can find a sequence $\{f_p\}_{p=1}^{\infty}$ in $C^{\infty}(M)$ such that

$$\begin{cases} f_p \longrightarrow f \text{ in } C(M), \\ \|f_p\|_{C^1(M)} \le \|f\|_{1,\infty}. \end{cases}$$

If $\varphi_{\varepsilon p}$ is a unique solution in $C^{\infty}(M)$ of the equation

$$(2.5) \qquad\qquad (T + \varepsilon\Lambda - \lambda I)\varphi_{\varepsilon p} = f_p \text{ on } M,$$

it follows from an application of inequality (2.4) that

$$\|\varphi_{\varepsilon p}\|_{C^1(M)}^2 \le C(\lambda)\|f_p\|_{C^1(M)}^2 \le C(\lambda)\|f\|_{1,\infty}^2.$$

This proves that the sequence $\{\varphi_{\varepsilon p}\}$ is uniformly bounded and equicontinuous on M. Hence, by virtue of the Ascoli-Arzelà theorem, one can choose a subsequence $\{\varphi_{\varepsilon' p'}\}$ which converges uniformly to a function $\varphi \in C(M)$, as $\varepsilon' \downarrow 0$ and $p' \to \infty$. Furthermore, since the unit ball in the Hilbert space $L^2(M)$ is *sequentially weakly compact* (cf. Yosida [18, Chapter V, Section 2, Theorem 1]), one may assume that the sequence $\{\partial_j \varphi_{\varepsilon' p'}\}$ converges weakly to a function ψ_j in $L^2(M)$, for each $1 \le j \le n$. Then we have

$$\partial_j \varphi = \psi_j \in L^2(M), \ 1 \le j \le n.$$

On the other hand, it is easy to verify that the set

$$K = \{g \in L^2(M) \, ; \, \|g\|_{\infty} \le \sqrt{C(\lambda)}\,\|f\|_{1,\infty}\}$$

is convex and strongly closed in $L^2(M)$. Thus it follows from an application of Mazur's theorem (cf. Yosida [18, Chapter V, Section 1, Theorem 11]) that the set K is *weakly closed* in $L^2(M)$. But we have

$$\begin{cases} \partial_j \varphi_{\varepsilon' p'} \in K, \\ \partial_j \varphi_{\varepsilon' p'} \longrightarrow \psi_j \text{ weakly in } L^2(M) \text{ for each } 1 \le j \le n. \end{cases}$$

Hence we find that

$$\partial_j \varphi = \psi_j \in K, \ 1 \le j \le n,$$

that is,

$$\|\partial_j \varphi\|_{\infty} \le \sqrt{C(\lambda)}\,\|f\|_{1,\infty}, \ 1 \le j \le n.$$

Summing up, we have proved that

$$\begin{cases} \varphi \in W^{1,\infty}(M), \\ \|\varphi\|_{1,\infty} \le C_1(\lambda)\|f\|_{1,\infty}, \end{cases}$$

where $C_1(\lambda) > 0$ is a constant *independent* of f.

Finally, by letting $\varepsilon' \downarrow 0$ and $p' \to \infty$ in the equation

$$(2.5') \qquad\qquad (T + \varepsilon'\Lambda - \lambda I)\varphi_{\varepsilon' p'} = f_{p'} \text{ on } M,$$

we obtain that

$$(T - \lambda I)\varphi = f \text{ on } M.$$

The proof of Claim I is complete.

3) Similarly, we can prove Theorem 2.1 for the spaces $W^{m,\infty}(M)$ where $m \ge 2$:

Claim II. *For each integer $m \geq 2$, there exists a constant $\lambda = \lambda(m) > 0$ such that for any $f \in W^{m,\infty}(M)$ one can find a function $\varphi \in W^{m,\infty}(M)$ satisfying*

$$(T - \lambda I)\varphi = f \text{ on } M,$$

and

$$\|\varphi\|_{m,\infty} \leq C_m(\lambda)\|f\|_{m,\infty}.$$

Here $C_m(\lambda) > 0$ is a constant independent of f.

4) Therefore, Theorem 2.1 follows from Claims I and II by a well-known interpolation argument, since the space $C^{k+\theta}(M)$ is a real interpolation space between the spaces $W^{k,\infty}(M)$ and $W^{k+1,\infty}(M)$:

$$C^{k+\theta}(M) = \left(W^{k,\infty}(M), W^{k+1,\infty}(M)\right)_{\theta,\infty}.$$

CHAPTER III

PROOF OF THEOREM 1

In this chapter we prove Theorem 1. First we reduce the problem of construction of Feller semigroups to the problem of *unique solvability* for the boundary value problem

$$\begin{cases} (\alpha - W)u = f & \text{in } D, \\ (\lambda - L)u = \varphi & \text{on } \partial D, \end{cases}$$

and then prove existence theorems for Feller semigroups. Here $\alpha > 0$ and $\lambda \geq 0$.

We explain the idea of the proof of Theorem 1.

(1) First we consider the following *Dirichlet problem*:

$$\begin{cases} (\alpha - W)v = f & \text{in } D, \\ v|_{\partial D} = 0 & \text{on } \partial D. \end{cases}$$

The existence and uniqueness theorem for this problem is well established in the framework of Hölder spaces. We let

$$v = G_\alpha^0 f.$$

The operator G_α^0 is the Green operator for the Dirichlet problem. Then it follows that a function u is a solution of the problem

$$(*) \qquad \begin{cases} (\alpha - W)u = f & \text{in } D, \\ Lu = 0 & \text{on } \partial D \end{cases}$$

if and only if the function $w = u - v$ is a solution of the problem

$$\begin{cases} (\alpha - W)w = 0 & \text{in } D, \\ Lw = -Lv = -LG_\alpha^0 f & \text{on } \partial D. \end{cases}$$

But we know that every solution w of the equation

$$(\alpha - W)w = 0 \ \text{ in } D$$

can be expressed by means of a single layer potential as follows:

$$w = H_\alpha \psi.$$

The operator H_α is the harmonic operator for the Dirichlet problem. Thus, by using the Green and harmonic operators, one can reduce the study of problem $(*)$ to that of the equation:

$$LH_\alpha \psi = -LG_\alpha^0 f.$$

It is known that the operator LH_α is a pseudo-differential operator of second order on the boundary ∂D.

(2) By using the Hölder space theory of pseudo-differential operators, we can show that if the boundary condition L is transversal on the boundary ∂D, then the operator LH_α is *bijective* in the framework of Hölder spaces. The crucial point in the proof is that we consider the term $\delta(Wu|_{\partial D})$ of viscosity in the boundary condition

$$Lu = L_0 u - \delta(Wu|_{\partial D})$$

as a term of "perturbation" of the boundary condition $L_0 u$.

Therefore, we find that a unique solution u of problem $(*)$ can be expressed as follows:

$$u = G_\alpha f = G_\alpha^0 f - H_\alpha \left(LH_\alpha^{-1} LG_\alpha^0 f \right).$$

This formula allows us to verify all the conditions of the generation theorems of Feller semigroups discussed in Chapter I, especially the *density* of the domain $D(\mathfrak{A})$ in the space $C(\bar{D})$.

3.1 General Existence Theorem for Feller Semigroups

The purpose of this section is to give a general existence theorem for Feller semigroups (Theorem 3.14) in terms of boundary value problems, following Section 9.6 of Taira [12] (cf. Bony-Courrège-Priouret [1]; Sato-Ueno [9]).

Let D be a bounded domain of Euclidean space $\mathbf{R}^N$, with smooth boundary ∂D; its closure $\bar{D} = D \cup \partial D$ is an N-dimensional, compact C^∞ manifold with boundary. We let W be a second-order *elliptic* Waldenfels operator with real coefficients such that

$$(0.1) \quad Wu(x) = Pu(x) + S_r u(x)$$

$$\equiv \left(\sum_{i,j=1}^{N} a^{ij}(x) \frac{\partial^2 u}{\partial x_i \partial x_j}(x) + \sum_{i=1}^{N} b^i(x) \frac{\partial u}{\partial x_i}(x) + c(x)u(x) \right)$$

$$+ \left(\int_D s(x,y) \left[u(y) - \sigma(x,y) \left(u(x) + \sum_{j=1}^{N} (y_j - x_j) \frac{\partial u}{\partial x_j}(x) \right) \right] dy \right),$$

where:

1) $a^{ij} \in C^\infty(\mathbf{R}^N)$, $a^{ij} = a^{ji}$ and there exists a constant $a_0 > 0$ such that

$$\sum_{i,j=1}^{N} a^{ij}(x)\xi_i\xi_j \geq a_0 |\xi|^2, \quad x \in \mathbf{R}^N, \ \xi \in \mathbf{R}^N.$$

2) $b^i \in C^\infty(\mathbf{R}^N)$.

3) $c \in C^\infty(\mathbf{R}^N)$ and $c \leq 0$ in D.

4) The integral kernel $s(x,y)$ is the distribution kernel of a properly supported, pseudo-differential operator $S \in L_{1,0}^{2-\kappa}(\mathbf{R}^N)$, $\kappa > 0$, which has the transmission property with respect to the boundary ∂D, and $s(x,y) \geq 0$ off the diagonal $\{(x,x); x \in \mathbf{R}^N\}$ in $\mathbf{R}^N \times \mathbf{R}^N$. The measure dy is the Lebesgue measure on $\mathbf{R}^N$.

5) The function $\sigma(x,y)$ is a C^∞ function on $\bar{D} \times \bar{D}$ such that $\sigma(x,y) = 1$ in a neighborhood of the diagonal $\{(x,x); x \in \bar{D}\}$ in $\bar{D} \times \bar{D}$. The function $\sigma(x,y)$ depends on the shape of the domain D. More precisely, it depends on a family of local charts on D in each of which the Taylor expansion is valid for functions u (cf. [1, Subsection I.2.7]). For example, if D is convex, then the Taylor expansion is valid in the whole D; so one may take $\sigma(x,y) \equiv 1$ on $\bar{D} \times \bar{D}$.

6) $W1(x) = c(x) + \int_D s(x,y)[1 - \sigma(x,y)]dy \le 0$ in D.

We remark that the integral operator

$$S_r : u(x) \longmapsto \int_D s(x,y)\left[u(y) - \sigma(x,y)\left(u(x) + \sum_{j=1}^{N}(y_j - x_j)\frac{\partial u}{\partial x_j}(x)\right)\right]dy$$

is a "regularization" of S, since the integrand is absolutely convergent. In fact, it suffices to note (cf. [4, Chapitre IV, Proposition 1]) that, for any compact $K \subset \mathbf{R}^N$, there exists a constant $C_K > 0$ such that the distribution kernel $s(x,y)$ of $S \in L^{2-\kappa}_{1,0}(\mathbf{R}^N)$, $\kappa > 0$, satisfies the estimate

$$|s(x,y)| \le \frac{C_K}{|x - y|^{N+2-\kappa}}, \quad x \in K, \, y \ne x.$$

Let L be a second-order Ventcel' boundary condition such that in local coordinates $(x_1, \cdots, x_{N-1})$
(0.2)
$$Lu(x') = Qu(x') + \mu(x')\frac{\partial u}{\partial \mathbf{n}}(x') - \delta(x')Wu(x') + \Gamma u(x')$$

$$\equiv \left(\sum_{i,j=1}^{N-1} \alpha^{ij}(x')\frac{\partial^2 u}{\partial x_i \partial x_j}(x') + \sum_{i=1}^{N-1}\beta^i(x')\frac{\partial u}{\partial x_i}(x') + \gamma(x')u(x')\right)$$

$$+ \mu(x')\frac{\partial u}{\partial \mathbf{n}}(x') - \delta(x')Wu(x') + \left(\eta(x')u(x') + \sum_{i=1}^{N-1}\zeta^i(x')\frac{\partial u}{\partial x_i}(x')\right.$$

$$+ \int_{\partial D} r(x',y')\left[u(y') - \tau(x',y')\left(u(x') + \sum_{j=1}^{N-1}(y_j - x_j)\frac{\partial u}{\partial x_j}(x')\right)\right]dy'$$

$$\left. + \int_D t(x',y)\left[u(y) - \tau(x',y)\left(u(x') + \sum_{j=1}^{N-1}(y_j - x_j)\frac{\partial u}{\partial x_j}(x')\right)\right]dy\right),$$

where:

1) The operator Q is a second-order degenerate elliptic differential operator on ∂D with non-positive principal symbol. In other words, the α^{ij} are the components of a C^∞ symmetric contravariant tensor of type $\binom{2}{0}$ on ∂D satisfying

$$\sum_{i,j=1}^{N-1} \alpha^{ij}(x')\xi_i\xi_j \ge 0, \quad x' \in \partial D, \, \xi' = \sum_{j=1}^{N-1}\xi_j dx_j \in T^*_{x'}(\partial D).$$

Here $T^*_{x'}(\partial D)$ is the cotangent space of ∂D at x'.

2) $Q1 = \gamma \in C^\infty(\partial D)$ and $\gamma \leq 0$ on ∂D.

3) $\mu \in C^\infty(\partial D)$ and $\mu \geq 0$ on ∂D.

4) $\delta \in C^\infty(\partial D)$ and $\delta \geq 0$ on ∂D.

5) $\mathbf{n} = (n_1, \ldots, n_N)$ is the unit interior normal to the boundary ∂D.

6) The integral kernel $r(x', y')$ is the distribution kernel of a pseudo-differential operator $R \in L_{1,0}^{2-\kappa_1}(\partial D)$, $\kappa_1 > 0$, and $r(x', y') \geq 0$ off the diagonal $\Delta_{\partial D} = \{(x', x'); x' \in \partial D\}$ in $\partial D \times \partial D$. The density dy' is a strictly positive density on ∂D.

7) The integral kernel $t(x, y)$ is the distribution kernel of a properly supported, pseudo-differential operator $T \in L_{1,0}^{2-\kappa_2}(\mathbf{R}^N)$, $\kappa_2 > 0$, which has the transmission property with respect to the boundary ∂D, and $t(x, y) \geq 0$ off the diagonal $\{(x, x); x \in \mathbf{R}^N\}$ in $\mathbf{R}^N \times \mathbf{R}^N$.

8) The function $\tau(x, y)$ is a C^∞ function on $\bar{D} \times \bar{D}$, with compact support in a neighborhood of the diagonal $\Delta_{\partial D}$, such that, at each point x' of ∂D, $\tau(x', y) = 1$ for y in a neighborhood of x' in $\bar{D}$. The function $\tau(x, y)$ depends on a family of local charts on $\bar{D}$ near ∂D in each of which the Taylor expansion is valid for functions u.

9) The operator Γ is a boundary condition of order $2 - \min(\kappa_1, \kappa_2)$, and satisfies the condition

$$\Gamma 1(x') = \eta(x') + \int_{\partial D} r(x', y')[1 - \tau(x', y')]dy'$$
$$+ \int_D t(x', y)[1 - \tau(x', y)]dy \leq 0 \text{ on } \partial D.$$

In this chapter, we are interested in the following problem:

Problem. *Given analytic data* (W, L), *can we construct a Feller semigroup* $\{T_t\}_{t \geq 0}$ *on* $\bar{D}$ *whose infinitesimal generator* $\mathfrak{A}$ *is characterized by* (W, L) ?

First we consider the following Dirichlet problem: For given functions f and φ defined in D and on ∂D, respectively, find a function u in D such that

$$(D) \qquad\qquad \begin{cases} (\alpha - W)u = f & \text{in } D, \\ u|_{\partial D} = \varphi & \text{on } \partial D, \end{cases}$$

where α is a positive number.

The next theorem summarizes the basic facts about the Dirichlet problem in the framework of *Hölder spaces* (cf. Bony-Courrège-Priouret [1, Théorème XV]):

Theorem 3.1. *Let* k *be an arbitrary non-negative integer and* $0 < \theta < 1$. *For any* $f \in C^{k+\theta}(\bar{D})$ *and any* $\varphi \in C^{k+2+\theta}(\partial D)$, *problem* (D) *has a unique solution* u *in* $C^{k+2+\theta}(\bar{D})$.

Theorem 3.1 with $k = 0$ tells us that problem (D) has a unique solution u in $C^{2+\theta}(\bar{D})$ for any $f \in C^\theta(\bar{D})$ and any $\varphi \in C^{2+\theta}(\partial D)$ $(0 < \theta < 1)$. Therefore we can introduce linear operators

$$G_\alpha^0 : C^\theta(\bar{D}) \longrightarrow C^{2+\theta}(\bar{D}),$$

and

$$H_\alpha : C^{2+\theta}(\partial D) \longrightarrow C^{2+\theta}(\bar{D})$$

as follows.

a) For any $f \in C^\theta(\bar{D})$, the function $G^0_\alpha f \in C^{2+\theta}(\bar{D})$ is the unique solution of the problem:

(3.1)
$$\begin{cases} (\alpha - W)G^0_\alpha f = f & \text{in } D, \\ G^0_\alpha f|_{\partial D} = 0 & \text{on } \partial D. \end{cases}$$

b) For any $\varphi \in C^{2+\theta}(\partial D)$, the function $H_\alpha \varphi \in C^{2+\theta}(\bar{D})$ is the unique solution of the problem:

(3.2)
$$\begin{cases} (\alpha - W)H_\alpha \varphi = 0 & \text{in } D, \\ H_\alpha \varphi|_{\partial D} = \varphi & \text{on } \partial D. \end{cases}$$

The operator G^0_α is called the *Green operator* and the operator H_α is called the *harmonic operator*, respectively.

Then we have the following results (cf. [12, Lemmas 9.6.2 and 9.6.3]):

Lemma 3.2. *The operator G^0_α $(\alpha > 0)$, considered from $C(\bar{D})$ into itself, is non-negative and continuous with norm*

$$\left\| G^0_\alpha \right\| = \left\| G^0_\alpha 1 \right\| = \max_{x \in \bar{D}} G^0_\alpha 1(x).$$

Lemma 3.3. *The operator H_α $(\alpha > 0)$, considered from $C(\partial D)$ into $C(\bar{D})$, is non-negative and continuous with norm*

$$\left\| H_\alpha \right\| = \left\| H_\alpha 1 \right\| = \max_{x \in \bar{D}} H_\alpha 1(x).$$

More precisely, we have the following (cf. [1, Proposition III.1.6]):

Theorem 3.4. *(i) (a) The operator G^0_α $(\alpha > 0)$ can be uniquely extended to a non-negative, bounded linear operator on $C(\bar{D})$ into itself, denoted again G^0_α, with norm*

$$\left\| G^0_\alpha \right\| = \left\| G^0_\alpha 1 \right\| \leq \frac{1}{\alpha}.$$

(b) For any $f \in C(\bar{D})$, we have

$$G^0_\alpha f \big|_{\partial D} = 0.$$

(c) For all $\alpha, \beta > 0$, the resolvent equation holds:

(3.3)
$$G^0_\alpha f - G^0_\beta f + (\alpha - \beta)G^0_\alpha G^0_\beta f = 0, \ f \in C(\bar{D}).$$

(d) For any $f \in C(\bar{D})$, we have

$$\lim_{\alpha \to +\infty} \alpha G^0_\alpha f(x) = f(x), \ x \in D.$$

Furthermore, if $f|_{\partial D} = 0$, then this convergence is uniform in $x \in \bar{D}$, that is,

$$\lim_{\alpha \to +\infty} \alpha G_\alpha^0 f = f \text{ in } C(\bar{D}).$$

(e) *The operator G_α^0 maps $C^{k+\theta}(\bar{D})$ into $C^{k+2+\theta}(\bar{D})$ for any non-negative integer k.*

(ii) (a') *The operator H_α ($\alpha > 0$) can be uniquely extended to a non-negative, bounded linear operator on $C(\partial D)$ into $C(\bar{D})$, denoted again H_α, with norm $\|H_\alpha\| \leq 1$.*

(b') *For any $\varphi \in C(\partial D)$, we have*

$$H_\alpha \varphi|_{\partial D} = \varphi.$$

(c') *For all $\alpha, \beta > 0$, we have*

(3.4) $$H_\alpha \varphi - H_\beta \varphi + (\alpha - \beta) G_\alpha^0 H_\beta \varphi = 0, \quad \varphi \in C(\partial D).$$

(d') *For any $\varphi \in C(\partial D)$, we have*

$$\lim_{\alpha \to +\infty} H_\alpha \varphi(x) = 0, \quad x \in D.$$

(e') *The operator H_α maps $C^{k+2+\theta}(\partial D)$ into $C^{k+2+\theta}(\bar{D})$ for any non-negative integer k.*

Now we consider the following boundary value problem $(*)$ in the framework of the spaces of *continuous functions*.

$(*)$ $$\begin{cases} (\alpha - W)u = f & \text{in } D, \\ Lu = 0 & \text{on } \partial D. \end{cases}$$

To do so, we introduce three operators associated with problem $(*)$.

(I) First we introduce a linear operator

$$W : C(\bar{D}) \longrightarrow C(\bar{D})$$

as follows.

(a) The domain $D(W)$ of W is the space $C^{2+\theta}(\bar{D})$.

(b) $Wu = Pu + S_r u$, $u \in D(W)$.

Then we have the following (cf. [12, Lemma 9.6.5]):

Lemma 3.5. *The operator W has its minimal closed extension $\overline{W}$ in the space $C(\bar{D})$.*

Remark 3.6. Since the injection: $C(\bar{D}) \longrightarrow \mathcal{D}'(D)$ is continuous, we have the formula:

$$\overline{W}u(x) = \sum_{i,j=1}^{N} a^{ij}(x) \frac{\partial^2 u}{\partial x_i \partial x_j}(x) + \sum_{i=1}^{N} b^i(x) \frac{\partial u}{\partial x_i}(x) + c(x)u(x)$$

$$+ \int_D s(x, y) \left[u(y) - \sigma(x, y) \left(u(x) + \sum_{j=1}^{N} (y_j - x_j) \frac{\partial u}{\partial x_j}(x) \right) \right] dy,$$

where the right-hand side is taken in the sense of *distributions*.

The extended operators $G_\alpha^0 : C(\bar{D}) \longrightarrow C(\bar{D})$ and $H_\alpha : C(\partial D) \longrightarrow C(\bar{D})$ ($\alpha > 0$) still satisfy formulas (3.1) and (3.2) respectively in the following sense (cf. [12, Lemma 9.6.7 and Corollary 9.6.8]):

Lemma 3.7. *(i) For any $f \in C(\bar{D})$, we have*

$$\begin{cases} G_\alpha^0 f \in D(\overline{W}), \\ (\alpha I - \overline{W}) G_\alpha^0 f = f \text{ in } D. \end{cases}$$

(ii) For any $\varphi \in C(\partial D)$, we have

$$\begin{cases} H_\alpha \varphi \in D(\overline{W}), \\ (\alpha I - \overline{W}) H_\alpha \varphi = 0 \text{ in } D. \end{cases}$$

Here $D(\overline{W})$ is the domain of the closed extension $\overline{W}$.

Corollary 3.8. *Every u in $D(\overline{W})$ can be written in the following form:*

$$(3.5) \qquad u = G_\alpha^0 \left((\alpha I - \overline{W}) u \right) + H_\alpha(u|_{\partial D}), \quad \alpha > 0.$$

(II) Secondly we introduce a linear operator

$$LG_\alpha^0 : C(\bar{D}) \longrightarrow C(\partial D)$$

as follows.

(a) The domain $D\left(LG_\alpha^0\right)$ of LG_α^0 is the space $C^\theta(\bar{D})$.

(b) $LG_\alpha^0 f = L\left(G_\alpha^0 f\right)$, $f \in D\left(LG_\alpha^0\right)$.

Then we have the following (cf. [12, Lemma 9.6.9]):

Lemma 3.9. *The operator LG_α^0 $(\alpha > 0)$ can be uniquely extended to a non-negative, bounded linear operator $\overline{LG_\alpha^0} : C(\bar{D}) \longrightarrow C(\partial D)$.*

The next lemma states a fundamental relationship between the operators $\overline{LG_\alpha^0}$ and $\overline{LG_\beta^0}$ for $\alpha, \beta > 0$ (cf. [12, Lemma 9.6.10]):

Lemma 3.10. *For any $f \in C(\bar{D})$, we have*

$$(3.6) \qquad \overline{LG_\alpha^0} f - \overline{LG_\beta^0} f + (\alpha - \beta) \overline{LG_\alpha^0} \, G_\beta^0 f = 0, \quad \alpha, \beta > 0.$$

(III) Finally we introduce a linear operator

$$LH_\alpha : C(\partial D) \longrightarrow C(\partial D)$$

as follows.

(a) The domain $D(LH_\alpha)$ of LH_α is the space $C^{2+\theta}(\partial D)$.

(b) $LH_\alpha \psi = L(H_\alpha \psi)$, $\psi \in D(LH_\alpha)$.

Then we have the following (cf. [12, Lemma 9.6.11]):

Lemma 3.11. *The operator LH_α $(\alpha > 0)$ has its minimal closed extension $\overline{LH_\alpha}$ in the space $C(\partial D)$.*

Remark 3.12. The operator $\overline{LH_\alpha}$ enjoys the following property:

$$(3.7) \qquad \text{If a function } \psi \text{ in the domain } D\left(\overline{LH_\alpha}\right) \text{ takes its } positive \text{ maximum}$$

at some point x' of ∂D, then we have

$$\overline{LH_\alpha} \psi(x') \leq 0.$$

The next lemma states a fundamental relationship between the operators $\overline{LH_\alpha}$ and $\overline{LH_\beta}$ for $\alpha, \beta > 0$ (cf. [12, Lemma 9.6.13]):

Lemma 3.13. *The domain* $D\left(\overline{LH_\alpha}\right)$ *of* $\overline{LH_\alpha}$ *does not depend on* $\alpha > 0$; *so we denote by* $\mathcal{D}$ *the common domain. Then we have*

$$(3.8) \qquad \overline{LH_\alpha}\psi - \overline{LH_\beta}\psi + (\alpha - \beta)\overline{LG_\alpha^0} H_\beta\psi = 0, \ \alpha,\beta > 0, \ \psi \in \mathcal{D}.$$

Now we can give a general existence theorem for Feller semigroups on ∂D in terms of boundary value problem $(*)$. The next theorem tells us that the operator $\overline{LH_\alpha}$ is the infinitesimal generator of some Feller semigroup on ∂D if and only if problem $(*)$ is solvable for sufficiently *many* functions φ in the space $C(\partial D)$ (cf. [12, Theorem 9.6.15]):

Theorem 3.14. *(i) If the operator* $\overline{LH_\alpha}$ ($\alpha > 0$) *is the infinitesimal generator of a Feller semigroup on* ∂D, *then, for each constant* $\lambda > 0$, *the boundary value problem*

$$(*') \qquad \begin{cases} (\alpha - W)u = 0 & \text{in } D, \\ (\lambda - L)u = \varphi & \text{on } \partial D \end{cases}$$

has a solution $u \in C^{2+\theta}(\bar{D})$ *for any* φ *in some dense subset of* $C(\partial D)$.

(ii) Conversely, if, for some constant $\lambda \geq 0$, *problem* $(*')$ *has a solution* $u \in C^{2+\theta}(\bar{D})$ *for any* φ *in some dense subset of* $C(\partial D)$, *then the operator* $\overline{LH_\alpha}$ *is the infinitesimal generator of some Feller semigroup on* ∂D.

We conclude this section by giving a precise meaning to the boundary conditions Lu for functions u in the domain $D(\overline{W})$.

We let

$$D(L) = \left\{ u \in D(\overline{W}) ; u|_{\partial D} \in \mathcal{D} \right\},$$

where $\mathcal{D}$ is the common domain of the operators $\overline{LH_\alpha}$, $\alpha > 0$. We remark that the domain $D(L)$ contains the space $C^{2+\theta}(\bar{D})$, since $C^{2+\theta}(\partial D) = D(LH_\alpha) \subset \mathcal{D}$. Corollary 3.8 tells us that every function u in $D(L) \subset D(\overline{W})$ can be written in the form:

$$(3.5) \qquad u = G_\alpha^0 \left((\alpha I - \overline{W})u\right) + H_\alpha(u|_{\partial D}), \ \alpha > 0.$$

Then we define

$$(3.9) \qquad Lu = \overline{LG_\alpha^0}\left((\alpha I - \overline{W})u\right) + \overline{LH_\alpha}(u|_{\partial D}).$$

The next lemma justifies definition (3.9) of Lu for all $u \in D(L)$ (cf. [12, Lemma 9.6.16]):

Lemma 3.15. *The right-hand side of formula (3.9) depends only on* u, *not on the choice of expression (3.5).*

3.2 Proof of Theorem 1

Now we consider the boundary condition:

$$Lu(x') = \sum_{i,j=1}^{N-1} \alpha^{ij}(x') \frac{\partial^2 u}{\partial x_i \partial x_j}(x') + \sum_{i=1}^{N-1} \beta^i(x') \frac{\partial u}{\partial x_i}(x') + \gamma(x')u(x')$$

$$+ \mu(x') \frac{\partial u}{\partial \mathbf{n}}(x') - \delta(x') Wu(x') + \eta(x')u(x') + \sum_{i=1}^{N-1} \zeta^i(x') \frac{\partial u}{\partial x_i}(x')$$

$$+ \int_{\partial D} r(x',y') \left[u(y') - \tau(x',y') \left(u(x') + \sum_{j=1}^{N-1} (y_j - x_j) \frac{\partial u}{\partial x_j}(x') \right) \right] dy'$$

$$+ \int_D t(x',y) \left[u(y) - \tau(x',y) \left(u(x') + \sum_{j=1}^{N-1} (y_j - x_j) \frac{\partial u}{\partial x_j}(x') \right) \right] dy.$$

We recall that the boundary condition L is said to be *transversal* on the boundary ∂D if it satisfies the condition

$$(3.10) \qquad \int_D t(x',y) dy = +\infty \quad \text{if } \mu(x') = \delta(x') = 0.$$

The next theorem proves Theorem 1:

Theorem 3.16. *We define a linear operator*

$$\mathfrak{A} : C(\bar{D}) \longrightarrow C(\bar{D})$$

as follows (cf. formula (0.3)).
   (a) *The domain* $D(\mathfrak{A})$ *of* $\mathfrak{A}$ *is the set*

$$(3.11) \qquad D(\mathfrak{A}) = \left\{ u \in D(\overline{W}) ; \ u|_{\partial D} \in \mathcal{D}, \ Lu = 0 \right\},$$

where $\mathcal{D}$ *is the common domain of the operators* $\overline{LH_\alpha}$, $\alpha > 0$.
   (b) $\mathfrak{A}u = \overline{W}u$, $u \in D(\mathfrak{A})$.
   If the boundary condition L is transversal on the boundary ∂D, then the operator $\mathfrak{A}$ is the infinitesimal generator of some Feller semigroup on $\bar{D}$, and the Green operator $G_\alpha = (\alpha I - \mathfrak{A})^{-1}$, $\alpha > 0$, is given by the following formula:

$$(3.12) \qquad G_\alpha f = G_\alpha^0 f - H_\alpha \left(\overline{LH_\alpha}^{-1} \left(\overline{LG_\alpha^0 f} \right) \right), \ f \in C(\bar{D}).$$

Remark 3.17. Intuitively, formula (3.12) asserts that if the boundary condition L is transversal on the boundary ∂D, then one can "piece together" a Markov process (Feller semigroup) on the boundary ∂D with W-diffusion in the interior D to construct a Markov process (Feller semigroup) on the closure $\bar{D} = D \cup \partial D$.

Proof of Theorem 3.16. We apply part (ii) of Theorem 1.3 to the operator $\mathfrak{A}$ defined by formula (3.11). The proof is divided into several steps.

1) We let

$$L_0 u(x') = \sum_{i,j=1}^{N-1} \alpha^{ij}(x') \frac{\partial^2 u}{\partial x_i \partial x_j}(x') + \sum_{i=1}^{N-1} \beta^i(x') \frac{\partial u}{\partial x_i}(x') + \gamma(x') u(x')$$

$$+ \mu(x') \frac{\partial u}{\partial \mathbf{n}}(x') + \eta(x') u(x') + \sum_{i=1}^{N-1} \zeta^i(x') \frac{\partial u}{\partial x_i}(x')$$

$$+ \int_{\partial D} r(x', y') \left[u(y') - \tau(x', y') \left(u(x') + \sum_{j=1}^{N-1} (y_j - x_j) \frac{\partial u}{\partial x_j}(x') \right) \right] dy'$$

$$+ \int_D t(x', y) \left[u(y) - \tau(x', y) \left(u(x') + \sum_{j=1}^{N-1} (y_j - x_j) \frac{\partial u}{\partial x_j}(x') \right) \right] dy \,,$$

and consider the term $-\delta(Wu|_{\partial D})$ in Lu as a term of "perturbation" of $L_0 u$:

$$Lu = L_0 u - \delta\left(Wu|_{\partial D}\right).$$

It is easy to see that the operator $\overline{LH_\alpha}$ can be decomposed as follows:

$$\overline{LH_\alpha} = \overline{L_0 H_\alpha} - \alpha \delta I.$$

First we prove that:

> The operator $\overline{L_0 H_\alpha}$ generates a Feller semigroup on ∂D,
> for *any* $\alpha > 0$.

To do so, we remark that

$$L_0 H_\alpha \varphi(x')$$

$$= \left(\sum_{i,j=1}^{N-1} \alpha^{ij}(x') \frac{\partial^2 \varphi}{\partial x_i \partial x_j}(x') + \sum_{i=1}^{N-1} \beta^i(x') \frac{\partial \varphi}{\partial x_i}(x') + \gamma(x') \varphi(x') \right)$$

$$+ \mu(x') \frac{\partial}{\partial \mathbf{n}}(H_\alpha \varphi)(x') + \left(\eta(x') \varphi(x') + \sum_{i=1}^{N-1} \zeta^i(x') \frac{\partial \varphi}{\partial x_i}(x') \right.$$

$$+ \int_{\partial D} r(x', y') \left[\varphi(y') - \tau(x', y') \left(\varphi(x') + \sum_{j=1}^{N-1} (y_j - x_j) \frac{\partial \varphi}{\partial x_j}(x') \right) \right] dy'$$

$$\left. + \int_D t(x', y) \left[H_\alpha \varphi(y) - \tau(x', y) \left(\varphi(x') + \sum_{j=1}^{N-1} (y_j - x_j) \frac{\partial \varphi}{\partial x_j}(x') \right) \right] dy \right).$$

But we have the following results:

a) The operator

$$\varphi(x') \longmapsto \sum_{i,j=1}^{N-1} \alpha^{ij}(x') \frac{\partial^2 \varphi}{\partial x_i \partial x_j}(x') + \sum_{i=1}^{N-1} \beta^i(x') \frac{\partial \varphi}{\partial x_i}(x') + \gamma(x') \varphi(x')$$

is a second-order degenerate elliptic differential operator on ∂D with non-positive principal symbol, and $\gamma(x') \leq 0$ on ∂D.

b) The operator

$$\varphi(x') \longmapsto \mu(x')\frac{\partial}{\partial \mathbf{n}}(H_\alpha\varphi)(x')$$

is a classical, pseudo-differential operator of order one on ∂D (cf. Hörmander [5], Seeley [10]).

c) The operator

$$\varphi(x') \longmapsto \int_{\partial D} r(x',y')\left[\varphi(y') - \tau(x',y')\left(\varphi(x') + \sum_{j=1}^{N-1}(y_j - x_j)\frac{\partial\varphi}{\partial x_j}(x')\right)\right]dy'$$

is a classical, pseudo-differential operator of order $2 - \kappa_1$ on ∂D.

d) The operator

$$\varphi(x') \longmapsto \int_D t(x',y)\left[H_\alpha\varphi(y) - \tau(x',y)\left(\varphi(x') + \sum_{j=1}^{N-1}(y_j - x_j)\frac{\partial\varphi}{\partial x_j}(x')\right)\right]dy$$

is a classical, pseudo-differential operator of order $2 - \kappa_2$ on ∂D, since the operator $T \in L_{1,0}^{2-\kappa_2}(\mathbf{R}^N)$ has the transmission property with respect to the boundary ∂D (cf. Boutet de Monvel [2]; Rempel-Schulze [8]).

e) Since the function $H_\alpha 1$ takes its positive maximum 1 only on the boundary ∂D, it follows from an application of the boundary point lemma (cf. Appendix, Lemma A.3) that

$$(3.13) \qquad \frac{\partial}{\partial \mathbf{n}}(H_\alpha 1)\bigg|_{\partial D} < 0 \quad \text{on } \partial D.$$

Hence we have

$$L_0 H_\alpha 1(x')$$
$$= \gamma(x') + \mu(x')\frac{\partial}{\partial \mathbf{n}}(H_\alpha 1)(x')$$
$$+ \left(\eta(x') + \int_{\partial D} r(x',y')[1 - \tau(x',y')]dy' + \int_D t(x',y)[1 - \tau(x',y)]dy\right)$$
$$+ \int_D t(x',y)[H_\alpha 1(y) - 1]dy$$
$$\leq 0 \quad \text{on } \partial D.$$

Thus, applying Theorem 2.1 to the operator $L_0 H_\alpha$, we find that:

$$(3.14) \qquad \text{If } \lambda > 0 \text{ is sufficiently large, then the range } R(L_0 H_\alpha - \lambda I)$$
$$\text{contains the space } C^{2+\theta}(\partial D).$$

This implies that the range $R(L_0 H_\alpha - \lambda I)$ is a *dense* subset of $C(\partial D)$. Therefore, applying part (ii) of Theorem 3.14 to the operator L_0, we obtain that the operator $\overline{L_0 H_\alpha}$ is the infinitesimal generator of some Feller semigroup on ∂D, for any $\alpha > 0$.

2) Next we prove that:

> The operator $\overline{LH_\alpha} = \overline{L_0 H_\alpha} - \alpha\delta I$ generates a Feller semigroup on ∂D, for *any* $\alpha > 0$.

We remark that the operator $-\alpha\delta I$ is a bounded linear operator on the space $C(\partial D)$ into itself, and satisfies condition (β') of Theorem 1.5, since $\alpha > 0$ and $\delta \geq 0$ on ∂D.

Therefore, applying Corollary 1.6 with

$$\mathfrak{A} = \overline{L_0 H_\alpha},$$
$$M = -\alpha\delta I,$$

we obtain that the operator $\overline{LH_\alpha} = \overline{L_0 H_\alpha} - \alpha\delta I$ is the infinitesimal generator of a Feller semigroup on ∂D, for any $\alpha > 0$.

3) Now we prove that:

(3.15)   The equation

$$\overline{LH_\alpha}\,\psi = \varphi$$

has a unique solution ψ in $D(\overline{LH_\alpha})$ for any $\varphi \in C(\partial D)$; hence the inverse $\overline{LH_\alpha}^{\,-1}$ of $\overline{LH_\alpha}$ can be defined on the whole space $C(\partial D)$.

Further the operator $-\overline{LH_\alpha}^{\,-1}$ is non-negative and bounded on $C(\partial D)$.

We have by inequality (3.13) and transversality condition (3.10)

$$LH_\alpha 1(x')$$
$$= \gamma(x') + \mu(x')\frac{\partial}{\partial \mathbf{n}}(H_\alpha 1)(x') - \alpha\delta(x')$$
$$+ \left(\eta(x') + \int_{\partial D} r(x', y')[1 - \tau(x', y')]dy' + \int_D t(x', y)[1 - \tau(x', y)]dy\right)$$
$$+ \int_D t(x', y)[H_\alpha 1(y) - 1]dy$$
$$< 0 \text{ on } \partial D,$$

and so

$$\ell_\alpha = -\sup_{x' \in \partial D} LH_\alpha 1(x') > 0.$$

Further, using Corollary 1.4 with $K = \partial D$, $\mathfrak{A} = \overline{LH_\alpha}$ and $c = \ell_\alpha$, we obtain that the operator $\overline{LH_\alpha} + \ell_\alpha I$ is the infinitesimal generator of some Feller semigroup on ∂D. Therefore, since $\ell_\alpha > 0$, it follows from an application of part (i) of Theorem 1.3 with $\mathfrak{A} = \overline{LH_\alpha} + \ell_\alpha I$ that the equation

$$-\overline{LH_\alpha}\,\psi = \left(\ell_\alpha I - (\overline{LH_\alpha} + \ell_\alpha I)\right)\psi = \varphi$$

has a unique solution $\psi \in D(\overline{LH_\alpha})$ for any $\varphi \in C(\partial D)$, and further that the operator $-\overline{LH_\alpha}^{-1} = \left(\ell_\alpha I - (\overline{LH_\alpha} + \ell_\alpha I)\right)^{-1}$ is non-negative and bounded on the space $C(\partial D)$ with norm

$$\left\|-\overline{LH_\alpha}^{-1}\right\| = \left\|\left(\ell_\alpha I - (\overline{LH_\alpha} + \ell_\alpha I)\right)^{-1}\right\| \leq \frac{1}{\ell_\alpha}.$$

4) By assertion (3.15), we can define the right-hand side of formula (3.12) for all $\alpha > 0$. We prove that:

(3.16) $$G_\alpha = (\alpha I - \mathfrak{A})^{-1}, \ \alpha > 0.$$

In view of Lemmas 3.7 and 3.13, it follows that we have for all $f \in C(\bar{D})$

$$\begin{cases} G_\alpha f = G_\alpha^0 f - H_\alpha \left(\overline{LH_\alpha}^{-1}\left(\overline{LG_\alpha^0 f}\right)\right) \in D(\overline{W}), \\ G_\alpha f|_{\partial D} = -\overline{LH_\alpha}^{-1}\left(\overline{LG_\alpha^0 f}\right) \in D\left(\overline{LH_\alpha}\right) = \mathcal{D}, \\ LG_\alpha f = \overline{LG_\alpha^0 f} - \overline{LH_\alpha}\left(\overline{LH_\alpha}^{-1}\left(\overline{LG_\alpha^0 f}\right)\right) = 0, \end{cases}$$

and that

$$(\alpha I - \overline{W})G_\alpha f = f.$$

This proves that

$$\begin{cases} G_\alpha f \in D(\mathfrak{A}), \\ (\alpha I - \mathfrak{A})G_\alpha f = f, \end{cases}$$

that is,

$$(\alpha I - \mathfrak{A})G_\alpha = I \ \text{ on } \ C(\bar{D}).$$

Therefore, in order to prove formula (3.16), it suffices to show the injectivity of the operator $\alpha I - \mathfrak{A}$ for $\alpha > 0$.

Assume that:

$$u \in D(\mathfrak{A}) \ \text{ and } \ (\alpha I - \mathfrak{A})u = 0.$$

Then, by Corollary 3.8, the function u can be written as

$$u = H_\alpha\left(u|_{\partial D}\right), \ u|_{\partial D} \in \mathcal{D} = D\left(\overline{LH_\alpha}\right).$$

Thus we have

$$\overline{LH_\alpha}\left(u|_{\partial D}\right) = Lu = 0.$$

In view of assertion (3.15), this implies that

$$u|_{\partial D} = 0,$$

so that

$$u = H_\alpha\left(u|_{\partial D}\right) = 0 \ \text{ in } \ D.$$

5) The non-negativity of G_α $(\alpha > 0)$ follows immediately from formula (3.12), since the operators G_α^0, H_α, $-\overline{LH_\alpha}^{-1}$ and $\overline{LG_\alpha^0}$ are all non-negative.

6) We prove that the operator G_α is bounded on the space $C(\bar{D})$ with norm

$$(3.17) \qquad\qquad \|G_\alpha\| \le \frac{1}{\alpha} \, , \ \alpha > 0.$$

To do so, it suffices to show that

$$(3.17') \qquad\qquad G_\alpha 1 \le \frac{1}{\alpha} \ \text{on} \ \bar{D}.$$

since G_α is non-negative on $C(\bar{D})$.

First it follows from the uniqueness property of solutions of problem (D') that

$$(3.18) \qquad\qquad \alpha G_\alpha^0 1 + H_\alpha 1 = 1 + G_\alpha^0(W1) \quad \text{on} \ \bar{D}.$$

In fact, the both sides have the same boundary value 1 and satisfy the same equation: $(\alpha - W)u = \alpha$ in D.

Applying the operator L to the both hand sides of equality (3.18), we obtain that

$$
\begin{aligned}
&- LH_\alpha 1(x') \\
&= -L1(x') - LG_\alpha^0(W1)(x') + \alpha LG_\alpha^0 1(x') \\
&= -\gamma(x') - \left(\eta(x') + \int_{\partial D} r(x',y')[1 - \tau(x',y')]dy' + \int_D t(x',y)[1 - \tau(x',y)]dy \right) \\
&\quad - \mu(x')\frac{\partial}{\partial \mathbf{n}}(G_\alpha^0(W1))(x') - \int_D t(x',y)G_\alpha^0(W1)(y)dy + \alpha LG_\alpha^0 1(x') \\
&\ge \alpha LG_\alpha^0 1(x') \quad \text{on} \ \partial D,
\end{aligned}
$$

since $G_\alpha^0(W1)|_{\partial D} = 0$ and $G_\alpha^0(W1) \le 0$ on $\bar{D}$. Hence we have by the non-negativity of $-\overline{LH_\alpha}^{-1}$

$$(3.19) \qquad\qquad -\overline{LH_\alpha}^{-1}\left(LG_\alpha^0 1 \right) \le \frac{1}{\alpha} \ \text{on} \ \partial D.$$

Using formula (3.12) with $f = 1$, inequality (3.19) and equality (3.18), we obtain that

$$
\begin{aligned}
G_\alpha 1 &= G_\alpha^0 1 + H_\alpha \left(-\overline{LH_\alpha}^{-1}\left(LG_\alpha^0 1 \right) \right) \\
&\le G_\alpha^0 1 + \frac{1}{\alpha} H_\alpha 1 \\
&= \frac{1}{\alpha} + \frac{1}{\alpha} G_\alpha^0(W1) \\
&\le \frac{1}{\alpha} \quad \text{on} \ \bar{D},
\end{aligned}
$$

since the operators H_α and G_α^0 are non-negative.

7) Finally we prove that:

$$(3.20) \qquad \text{The domain } D(\mathfrak{A}) \text{ is everywhere } dense \text{ in the space } C(\bar{D}).$$

7-1) Before the proof, we need some lemmas on the behavior of G_α^0, H_α and $-\overline{LH_\alpha}^{-1}$ as $\alpha \to +\infty$ (cf. [1, Proposition III.1.6]; [12, Lemmas 9.6.19 and 9.6.20]):

Lemma 3.18. *For all* $f \in C(\bar{D})$, *we have*

(3.21) $$\lim_{\alpha \to +\infty} [\alpha G_\alpha^0 f + H_\alpha (f|_{\partial D})] = f \quad in \; C(\bar{D}).$$

Lemma 3.19. *The function*

$$\frac{\partial}{\partial \mathbf{n}} (H_\alpha 1)\Big|_{\partial D}$$

diverges to $-\infty$ *uniformly and monotonically as* $\alpha \to +\infty$.

Corollary 3.20. *If the boundary condition* L *is transversal on the boundary* ∂D, *then we have*

$$\lim_{\alpha \to +\infty} \| - \overline{LH_\alpha}^{-1} \| = 0.$$

Proof. We recall that

$$LH_\alpha 1(x')$$

$$= \gamma(x') + \mu(x') \frac{\partial}{\partial \mathbf{n}} (H_\alpha 1)(x') - \alpha \delta(x')$$

$$+ \left(\eta(x') + \int_{\partial D} r(x', y')[1 - \tau(x', y')] dy' + \int_D t(x', y)[1 - \tau(x', y)] dy \right)$$

$$+ \int_D t(x', y)[H_\alpha 1(y) - 1] dy$$

$$\leq \mu(x') \frac{\partial}{\partial \mathbf{n}} (H_\alpha 1)(x') - \alpha \delta(x') + \int_D t(x', y)[H_\alpha 1(y) - 1] dy.$$

But it follows from an application of Beppo-Levi's theorem that

$$\lim_{\alpha \to +\infty} \int_D t(x', y)[H_\alpha 1(y) - 1] dy = - \int_D t(x', y) dy,$$

since the function $H_\alpha 1$ converges to zero in D monotonically as $\alpha \to +\infty$.

Hence we obtain from Lemma 3.19 that if the boundary condition L is transversal on the boundary ∂D, that is, if we have

$$\int_D t(x', y) dy = +\infty \quad \text{if } \mu(x') = \delta(x') = 0,$$

then the function $LH_\alpha 1$ diverges to $-\infty$ monotonically as $\alpha \to +\infty$. By Dini's theorem, this convergence is uniform in $x' \in \partial D$. Hence the function

$$\frac{1}{LH_\alpha 1(x')}$$

converges to zero uniformly in $x' \in \partial D$ as $\alpha \to +\infty$. This gives that

$$\left\| -\overline{LH_\alpha}^{-1} \right\| = \left\| -\overline{LH_\alpha}^{-1} 1 \right\|$$

$$\leq \left\|\frac{1}{LH_\alpha 1}\right\| \longrightarrow 0 \quad \text{as } \alpha \to +\infty,$$

since we have

$$1 = \frac{-LH_\alpha 1(x')}{|LH_\alpha 1(x')|} \leq \left\|\frac{1}{LH_\alpha 1}\right\| (-LH_\alpha 1(x')), \quad x' \in \partial D.$$

7-2) *Proof of assertion (3.20)*

In view of formula (3.16) and inequality (3.17), it suffices to prove that

$$(3.22) \qquad \lim_{\alpha \to +\infty} \|\alpha G_\alpha f - f\| = 0, \ f \in C^{2+\theta}(\bar{D}),$$

since the space $C^{2+\theta}(\bar{D})$ is dense in $C(\bar{D})$.

First we remark that

$$\begin{aligned}
\|\alpha G_\alpha f - f\| &= \left\| \alpha G_\alpha^0 f - \alpha H_\alpha \left(\overline{LH_\alpha}^{-1} \left(LG_\alpha^0 f \right) \right) - f \right\| \\
&\leq \left\| \alpha G_\alpha^0 f + H_\alpha \left(f|_{\partial D} \right) - f \right\| \\
&\quad + \left\| -\alpha H_\alpha \left(\overline{LH_\alpha}^{-1} \left(LG_\alpha^0 f \right) \right) - H_\alpha \left(f|_{\partial D} \right) \right\| \\
&\leq \left\| \alpha G_\alpha^0 f + H_\alpha \left(f|_{\partial D} \right) - f \right\| \\
&\quad + \left\| -\alpha \overline{LH_\alpha}^{-1} \left(LG_\alpha^0 f \right) - f|_{\partial D} \right\|.
\end{aligned}$$

Thus, in view of formula (3.21), it suffices to show that

$$(3.23) \qquad \lim_{\alpha \to +\infty} \left[-\alpha \overline{LH_\alpha}^{-1} \left(LG_\alpha^0 f \right) - f|_{\partial D} \right] = 0 \quad \text{in } C(\partial D).$$

Take a constant β such that $0 < \beta < \alpha$, and write

$$f = G_\beta^0 g + H_\beta \varphi,$$

where (cf. formula (3.5)):

$$\begin{cases} g = (\beta - W)f \in C^\theta(\bar{D}), \\ \varphi = f|_{\partial D} \in C^{2+\theta}(\partial D). \end{cases}$$

Then, using equations (3.3) (with $f = g$) and (3.4), we obtain that

$$\begin{aligned}
G_\alpha^0 f &= G_\alpha^0 G_\beta^0 g + G_\alpha^0 H_\beta \varphi \\
&= \frac{1}{\alpha - \beta} \left(G_\beta^0 g - G_\alpha^0 g + H_\beta \varphi - H_\alpha \varphi \right).
\end{aligned}$$

Hence we have

$$\left\| -\alpha \overline{LH_\alpha}^{-1} \left(LG_\alpha^0 f \right) - f|_{\partial D} \right\|$$

$$= \left\| \frac{\alpha}{\alpha - \beta} \left(-\overline{LH_\alpha}^{-1} \right) \left(LG_\beta^0 g - LG_\alpha^0 g + LH_\beta \varphi \right) + \frac{\alpha}{\alpha - \beta} \varphi - \varphi \right\|$$

$$\leq \frac{\alpha}{\alpha - \beta} \left\| -\overline{LH_\alpha}^{-1} \right\| \cdot \left\| LG_\beta^0 g + LH_\beta \varphi \right\|$$

$$+ \frac{\alpha}{\alpha - \beta} \left\| -\overline{LH_\alpha}^{-1} \right\| \cdot \left\| LG_\alpha^0 \right\| \cdot \|g\| + \frac{\beta}{\alpha - \beta} \|\varphi\|.$$

By Corollary 3.20, it follows that the first term on the last inequality converges to zero as $\alpha \to +\infty$. For the second term, using formula (3.6) with $f = 1$ and the non-negativity of G_β^0 and LG_α^0, we find that

$$\begin{aligned} \|LG_\alpha^0\| &= \|LG_\alpha^0 1\| \\ &= \|LG_\beta^0 1 - (\alpha - \beta)LG_\alpha^0 G_\beta^0 1\| \\ &\leq \|LG_\beta^0 1\|. \end{aligned}$$

Hence the second term also converges to zero as $\alpha \to +\infty$. It is clear that the third term converges to zero as $\alpha \to +\infty$. This completes the proof of assertion (3.23) and hence that of assertion (3.22).

8) Summing up, we have proved that the operator $\mathfrak{A}$, defined by formula (3.11), satisfies conditions (a) through (d) in Theorem 1.3. Hence it follows from an application of the same theorem that the operator $\mathfrak{A}$ is the infinitesimal generator of some Feller semigroup on $\bar{D}$.

The proof of Theorem 3.16 and hence that of Theorem 1 is now complete.

CHAPTER IV

PROOF OF THEOREM 2

The idea of our approach is stated as follows.

1) Assume that condition (H) is satisfied:

(H) There exists a *transversal* Ventcel' boundary condition L_ν of second order such that

$$Lu(x') = \mu(x')L_\nu u(x') + \gamma(x')u(x') - \delta(x')Wu(x'), \quad x' \in \partial D.$$

First we consider the boundary value problem

$$\begin{cases} (\alpha - W)v = f & \text{in } D, \\ L_\nu v = 0 & \text{on } \partial D, \end{cases}$$

and let

$$v = G_\alpha^\nu f.$$

Then it is easy to see that a function u is a solution of the problem

$$(**) \qquad \begin{cases} (\alpha - W)u = f & \text{in } D, \\ Lu = \mu L_\nu u + \gamma u - \delta(Wu) = 0 & \text{on } \partial D \end{cases}$$

if and only if there exists a solution ψ of the equation

$$LH_\alpha \psi = (\alpha\delta - \gamma)\left(G_\alpha^\nu f|_{\partial D}\right) - \delta(f|_{\partial D}).$$

2) Next we show that if the condition

(A) $\mu(x') + \delta(x') - \gamma(x') > 0$ on ∂D

is satisfied, then the operator LH_α is *bijective* in the framework of Hölder spaces. This is proved by using the Hölder space theory of pseudo-differential operators, just as in the proof of Theorem 1.

3) Therefore, we find that a unique solution u of problem $(**)$ can be expressed as follows:

$$u = G_\alpha f = G_\alpha^\nu f - H_\alpha\left(\overline{LH_\alpha}^{-1}\left(LG_\alpha^\nu f\right)\right).$$

This formula allows us to verify all the conditions of the generation theorems of Feller semigroups discussed in Section 1.2.

4.1 The Space $C_0(\bar{D} \backslash M)$

First we consider a one-point compactification $K_\partial = K \cup \{\partial\}$ of the space $K = \bar{D} \backslash M$, where

$$M = \{x' \in \partial D; \mu(x') = \delta(x') = 0\}.$$

We say that two points x and y of $\bar{D}$ are equivalent modulo M if $x = y$ or x, $y \in M$; we then write $x \sim y$. It is easy to verify that this relation $\sim$ enjoys the so-called equivalence laws. We denote by $\bar{D}/M$ the totality of equivalence classes modulo M. On the set $\bar{D}/M$ we define the quotient topology induced by the projection

$$q : \bar{D} \longrightarrow \bar{D}/M.$$

That is, a subset O of $\bar{D}/M$ is defined to be open if and only if the inverse image $q^{-1}(O)$ of O is open in $\bar{D}$. It is easy to see that the topological space $\bar{D}/M$ is a *one-point compactification* of the space $\bar{D} \backslash M$ and that the *point at infinity* ∂ corresponds to the set M:

$$K_\partial = \bar{D}/M,$$
$$\partial = M.$$

Further we find that:

(i) If $\tilde{u}$ is a continuous function defined on K_∂, then the function $\tilde{u} \circ q$ is continuous on $\bar{D}$ and constant on M.

(ii) Conversely, if u is a continuous function defined on $\bar{D}$ and constant on M, then it defines a continuous function $\tilde{u}$ on K_∂.

In other words, we have the following isomorphism:

$$(4.1) \qquad C(K_\partial) \cong \left\{ u \in C(\bar{D}); u \text{ is constant on } M \right\}.$$

Now we introduce a closed subspace of $C(K_\partial)$ as in Section 1.1:

$$C_0(K) = \left\{ u \in C(K_\partial); u(\partial) = 0 \right\}.$$

Then we have by assertion (4.1)

$$(4.2) \qquad C_0(K) \cong C_0(\bar{D} \backslash M) = \left\{ u \in C(\bar{D}); u = 0 \text{ on } M \right\}.$$

4.2 Proof of Theorem 2

We shall apply part (ii) of Theorem 1.3 to the operator $\mathfrak{A}$ defined by formula (0.4).
First we simplify the boundary condition:

$$Lu = 0 \quad \text{on } \partial D.$$

Assume that conditions (A) and (H) are satisfied:
(A) $\mu(x') + \delta(x') - \gamma(x') > 0$ on ∂D.
(H) There exists a *transversal* Ventcel' boundary condition L_ν of second order such that

$$Lu(x') = \mu(x')L_\nu u(x') + \gamma(x')u(x') - \delta(x')Wu(x'), \quad x' \in \partial D,$$

where the boundary condition L_ν is given in local coordinates $(x_1, \cdots, x_{N-1})$ by the formula

$$
\begin{aligned}
L_\nu u(x') =& \sum_{i,j=1}^{N-1} \bar{\alpha}^{ij}(x') \frac{\partial^2 u}{\partial x_i \partial x_j}(x') + \sum_{i=1}^{N-1} \bar{\beta}^i(x') \frac{\partial u}{\partial x_i}(x') \\
&+ \frac{\partial u}{\partial \mathbf{n}}(x') + \bar{\eta}(x')u(x') + \sum_{i=1}^{N-1} \bar{\zeta}^i(x') \frac{\partial u}{\partial x_i}(x') \\
&+ \int_{\partial D} \bar{r}(x', y') \left[u(y') - \tau(x', y') \left(u(x') + \sum_{j=1}^{N-1} (y_j - x_j) \frac{\partial u}{\partial x_j}(x') \right) \right] dy' \\
&+ \int_D \bar{t}(x', y) \left[u(y) - \tau(x', y) \left(u(x') + \sum_{j=1}^{N-1} (y_j - x_j) \frac{\partial u}{\partial x_j}(x') \right) \right] dy,
\end{aligned}
$$

and satisfies the condition

$$\bar{\eta}(x') + \int_{\partial D} \bar{r}(x', y')[1 - \tau(x', y')]dy' + \int_D \bar{t}(x', y)[1 - \tau(x', y)]dy \leq 0 \quad \text{on } \partial D.$$

Then we find that the boundary condition

$$Lu = \mu L_\nu u + \gamma(u|_{\partial D}) - \delta(Wu|_{\partial D}) = 0 \quad \text{on } \partial D$$

is equivalent to the condition:

$$\left(\frac{\mu}{\mu + \delta - \gamma} \right) L_\nu u + \left(\frac{\gamma}{\mu + \delta - \gamma} \right) (u|_{\partial D}) - \left(\frac{\delta}{\mu + \delta - \gamma} \right) (Wu|_{\partial D}) = 0 \text{ on } \partial D.$$

If we let

$$\tilde{\mu} = \frac{\mu}{\mu + \delta - \gamma}, \quad \tilde{\gamma} = \frac{\gamma}{\mu + \delta - \gamma}, \quad \tilde{\delta} = \frac{\delta}{\mu + \delta - \gamma},$$

then we have

$$\tilde{\mu} L_\nu u + \tilde{\gamma}(u|_{\partial D}) - \tilde{\delta}(Wu|_{\partial D}) = 0 \quad \text{on } \partial D$$

and

$$\begin{cases} 0 \leq \tilde{\mu} \leq 1 & \text{on } \partial D, \\ 0 \leq \tilde{\delta} \leq 1 & \text{on } \partial D, \\ 0 \leq \tilde{\mu} + \tilde{\delta} \leq 1 & \text{on } \partial D, \\ \tilde{\gamma} = \tilde{\mu} + \tilde{\delta} - 1 & \text{on } \partial D. \end{cases}$$

Therefore, one may assume that the boundary condition L is of the form:

$$(4.3) \qquad Lu = \mu L_\nu u + (\mu + \delta - 1)(u|_{\partial D}) - \delta(Wu|_{\partial D}),$$

with

$$\begin{cases} 0 \leq \mu \leq 1 & \text{on } \partial D, \\ 0 \leq \delta \leq 1 & \text{on } \partial D, \\ 0 \leq \mu + \delta \leq 1 & \text{on } \partial D. \end{cases}$$

Furthermore, we remark that

$$\overline{LG_\alpha^0}f = \mu \overline{L_\nu G_\alpha^0}f + \delta(f|_{\partial D}),$$

and

$$\overline{LH_\alpha}\varphi = \mu \overline{L_\nu H_\alpha}\varphi + (\mu + \delta - 1)\varphi - \alpha\delta\varphi.$$

Hence, in view of definition (3.9), it follows that

$$(4.3') \qquad Lu = \mu L_\nu u + (\mu + \delta - 1)(u|_{\partial D}) - \delta(\overline{W}u|_{\partial D}), \ u \in D(L).$$

Therefore, the next theorem proves Theorem 2:

Theorem 4.1. *We define a linear operator*

$$\mathfrak{A} : C_0(\bar{D}\backslash M) \longrightarrow C_0(\bar{D}\backslash M)$$

as follows (cf. formula (3.11)).
(a) The domain $D(\mathfrak{A})$ of $\mathfrak{A}$ is the set

$$(4.4) \qquad D(\mathfrak{A}) = \{u \in C_0(\bar{D}\backslash M) ; \overline{W}u \in C_0(\bar{D}\backslash M),$$
$$Lu = \mu L_\nu u + (\mu + \delta - 1)(u|_{\partial D}) - \delta(\overline{W}u|_{\partial D}) = 0\}.$$

(b) $\mathfrak{A}u = \overline{W}u, \ u \in D(\mathfrak{A}).$
Assume that the following condition (A') is satisfied:
(A') $0 \leq \mu(x') + \delta(x') \leq 1$ on ∂D.
Then the operator $\mathfrak{A}$ is the infinitesimal generator of some Feller semigroup $\{T_t\}_{t\geq 0}$ *on $\bar{D}\backslash M$, and the Green operator $G_\alpha = (\alpha I - \mathfrak{A})^{-1}$, $\alpha > 0$, is given by the following formula:*

$$(4.5) \qquad G_\alpha f = G_\alpha^\nu f - H_\alpha\left(\overline{LH_\alpha}^{-1}(LG_\alpha^\nu f)\right), \ f \in C_0(\bar{D}\backslash M).$$

Here G_α^ν is the Green operator for the boundary condition L_ν given by formula (3.12):

$$G_\alpha^\nu f = G_\alpha^0 f - H_\alpha \left(\overline{L_\nu H_\alpha}^{-1} \left(\overline{L_\nu G_\alpha^0} f \right) \right), \quad f \in C(\bar{D}).$$

Proof. We apply part (ii) of Theorem 1.3 to the operator $\mathfrak{A}$ defined by formula (4.4), just as in the proof of Theorem 1. The proof is divided into several steps.

1) We let

$$L_1 u = \mu L_\nu u + (\mu + \delta - 1)(u|_{\partial D}),$$

and consider the term $-\delta(W u|_{\partial D})$ in Lu as a term of "perturbation" of $L_1 u$:

$$Lu = L_1 u - \delta (W u|_{\partial D}).$$

It is easy to see that the operator $\overline{LH_\alpha}$ can be decomposed as follows:

$$\overline{LH_\alpha} = \overline{L_1 H_\alpha} - \alpha \delta I.$$

First we prove that:

> The operator $\overline{L_1 H_\alpha}$ generates a Feller semigroup on ∂D,
>
> for *any* $\alpha > 0$.

To do so, we remark that

$$L_1 H_\alpha \varphi(x') = \mu(x') L_\nu H_\alpha \varphi(x') + (\mu(x') + \delta(x') - 1)\varphi(x'), \quad x' \in \partial D,$$

and

$$
\begin{aligned}
L_1 & H_\alpha 1(x') \\
&= \mu(x') \frac{\partial}{\partial \mathbf{n}} (H_\alpha 1)(x') + (\mu(x') + \delta(x') - 1) \\
&\quad + \mu(x') \left(\bar{\eta}(x') + \int_{\partial D} \bar{r}(x', y')[1 - \tau(x', y')] dy' + \int_D \bar{t}(x', y)[1 - \tau(x', y)] dy \right) \\
&\quad + \mu(x') \int_D \bar{t}(x', y)[H_\alpha 1(y) - 1] dy \\
&\leq 0 \text{ on } \partial D.
\end{aligned}
$$

Thus, applying Theorem 2.1 to the operator $L_1 H_\alpha$ (cf. the proof of assertion (3.14)), we obtain that:

(4.6)   If $\lambda > 0$ is sufficiently large, then the range $R(L_1 H_\alpha - \lambda I)$

contains the space $C^{2+\theta}(\partial D)$.

This implies that the range $R(L_1 H_\alpha - \lambda I)$ is a *dense* subset of $C(\partial D)$. Therefore, applying part (ii) of Theorem 3.14 to the operator L_1, we obtain that the operator $\overline{L_1 H_\alpha}$ is the infinitesimal generator of some Feller semigroup on ∂D, for any $\alpha > 0$.

2) Next we prove that:

> The operator $\overline{LH_\alpha}$ generates a Feller semigroup on ∂D,
> for *any* $\alpha > 0$.

We remark that the operator $-\alpha\delta I$ is a bounded linear operator on the space $C(\partial D)$ into itself, and satisfies condition (β') of Theorem 1.5, since $\alpha > 0$ and $\delta \geq 0$ on ∂D.

Therefore, applying Corollary 1.6 with

$$\mathfrak{A} = \overline{L_1 H_\alpha},$$
$$M = -\alpha\delta I,$$

we obtain that the operator $\overline{LH_\alpha} = \overline{L_1 H_\alpha} - \alpha\delta I$ is the infinitesimal generator of some Feller semigroup on ∂D, for any $\alpha > 0$.

3) Now we prove that:

(4.7)   If condition (A') is satisfied, then the equation

$$\overline{LH_\alpha}\,\psi = \varphi$$

has a unique solution ψ in $D\left(\overline{LH_\alpha}\right)$ for any $\varphi \in C(\partial D)$; hence the inverse $\overline{LH_\alpha}^{-1}$ of $\overline{LH_\alpha}$ can be defined on the whole space $C(\partial D)$.

Further the operator $-\overline{LH_\alpha}^{-1}$ is non-negative and bounded on $C(\partial D)$.

Since we have by inequality (3.13) and condition (A')

$$
\begin{aligned}
LH_\alpha 1(x') \\
= \mu(x')L_\nu H_\alpha 1(x') &+ (\mu(x') + \delta(x') - 1) - \alpha\delta(x') \\
= \mu(x')\frac{\partial}{\partial\mathbf{n}}(H_\alpha 1)(x') &+ (\mu(x') + \delta(x') - 1) - \alpha\delta(x') \\
&+ \mu(x')\left(\bar{\eta}(x') + \int_{\partial D}\bar{r}(x',y')[1 - \tau(x',y')]dy' + \int_D \bar{t}(x',y)[1 - \tau(x',y)]dy\right) \\
&+ \mu(x')\int_D \bar{t}(x',y)[H_\alpha 1(y) - 1]dy \\
< 0 \ \text{on}\ \partial D,
\end{aligned}
$$

it follows that

$$k_\alpha = -\sup_{x' \in \partial D} LH_\alpha 1(x') > 0.$$

Here we remark that the constants k_α are increasing in $\alpha > 0$:

$$\alpha \geq \beta > 0 \implies k_\alpha \geq k_\beta.$$

Further, using Corollary 1.4 with $K = \partial D$, $\mathfrak{A} = \overline{LH_\alpha}$ and $c = k_\alpha$, we obtain that the operator $\overline{LH_\alpha} + k_\alpha I$ is the infinitesimal generator of some Feller semigroup on

∂D. Therefore, since $k_\alpha > 0$, it follows from an application of part (i) of Theorem 1.3 with $\mathfrak{A} = \overline{LH_\alpha} + k_\alpha I$ that the equation

$$-\overline{LH_\alpha}\,\psi = \left(k_\alpha I - (\overline{LH_\alpha} + k_\alpha I)\right)\psi = \varphi$$

has a unique solution $\psi \in D\left(\overline{LH_\alpha}\right)$ for any $\varphi \in C(\partial D)$, and further that the operator $-\overline{LH_\alpha}^{-1} = \left(k_\alpha I - (\overline{LH_\alpha} + k_\alpha I)\right)^{-1}$ is non-negative and bounded on the space $C(\partial D)$ with norm

$$(4.8) \qquad \left\|-\overline{LH_\alpha}^{-1}\right\| = \left\|\left(k_\alpha I - (\overline{LH_\alpha} + k_\alpha I)\right)^{-1}\right\| \le \frac{1}{k_\alpha}\,.$$

4) By assertion (4.7), we can define the operator G_α by formula (4.5) for all $\alpha > 0$. We prove that:

$$(4.9) \qquad G_\alpha = (\alpha I - \mathfrak{A})^{-1}\,, \quad \alpha > 0.$$

By virtue of Lemma 3.7 and Theorem 3.16, it follows that we have for all $f \in C_0(\bar{D}\backslash M)$

$$G_\alpha f \in D(\overline{W}),$$

and

$$\overline{W}G_\alpha f = \alpha G_\alpha f - f.$$

Further we have

$$(4.10) \qquad LG_\alpha f = LG_\alpha^\nu f - \overline{LH_\alpha}\left(\overline{LH_\alpha}^{-1}\left(LG_\alpha^\nu f\right)\right) = 0 \quad \text{on } \partial D.$$

But we recall that

$$(4.3') \qquad Lu = \mu L_\nu u + (\mu + \delta - 1)(u|_{\partial D}) - \delta\left(\overline{W}u|_{\partial D}\right)\,, \quad u \in D(L).$$

Hence formula (4.10) is equivalent to the following:

$$(4.10') \qquad \mu L_\nu(G_\alpha f) + (\mu + \delta - 1)(G_\alpha f|_{\partial D}) - \delta((\alpha G_\alpha f - f)|_{\partial D}) = 0 \quad \text{on } \partial D.$$

This implies that

$$G_\alpha f = 0 \quad \text{on } M = \{x' \in \partial D\,;\,\mu(x') = \delta(x') = 0\},$$

and so

$$\overline{W}G_\alpha f = \alpha G_\alpha f - f = 0 \quad \text{on } M.$$

Summing up, we have proved that

$$G_\alpha f \in D(\mathfrak{A}) = \left\{u \in C_0(\bar{D}\backslash M)\,;\,\overline{W}u \in C_0(\bar{D}\backslash M),\,Lu = 0\right\},$$

and

$$(\alpha I - \mathfrak{A})G_\alpha f = f,\quad f \in C_0(\bar{D}\backslash M),$$

that is,

$$(\alpha I - \mathfrak{A})G_\alpha = I \quad \text{on} \ \ C_0(\bar{D} \backslash M).$$

Therefore, in order to prove formula (4.9), it suffices to show the injectivity of the operator $\alpha I - \mathfrak{A}$ for $\alpha > 0$.

Assume that

$$u \in D(\mathfrak{A}) \ \ \text{and} \ \ (\alpha I - \mathfrak{A})u = 0.$$

Then, by Corollary 3.8, the function u can be written as

$$u = H_\alpha \left(u|_{\partial D} \right), \ \ u|_{\partial D} \in \mathcal{D} = D \left(\overline{LH_\alpha} \right).$$

Thus we have

$$\overline{LH_\alpha} \left(u|_{\partial D} \right) = Lu = 0.$$

In view of assertion (4.7), this implies that

$$u|_{\partial D} = 0,$$

so that

$$u = H_\alpha \left(u|_{\partial D} \right) = 0 \ \ \text{in} \ D.$$

5) Now we prove the following assertions:

(i) The operator G_α is non-negative on the space $C_0(\bar{D} \backslash M)$:

$$f \in C_0(\bar{D} \backslash M), \, f \geq 0 \ \text{on} \ \bar{D} \backslash M \implies G_\alpha f \geq 0 \ \text{on} \ \bar{D} \backslash M.$$

(ii) The operator G_α is bounded on the space $C_0(\bar{D} \backslash M)$ with norm

$$\|G_\alpha\| \leq \frac{1}{\alpha}, \ \alpha > 0.$$

(iii) The domain $D(\mathfrak{A})$ is everywhere dense in the space $C_0(\bar{D} \backslash M)$.

i) We show the non-negativity of G_α on the space $C(\bar{D})$:

$$f \in C(\bar{D}), \, f \geq 0 \ \text{on} \ \bar{D} \implies G_\alpha f \geq 0 \ \text{on} \ \bar{D}.$$

Recall that the Dirichlet problem

$$(D') \qquad\qquad \begin{cases} (\alpha - W)u = f & \text{in} \ D, \\ u|_{\partial D} = \varphi & \text{on} \ \partial D \end{cases}$$

is uniquely solvable. Hence it follows that

$$(4.11) \qquad\qquad G_\alpha^\nu f = H_\alpha \left(G_\alpha^\nu f|_{\partial D} \right) + G_\alpha^0 f \quad \text{on} \ \bar{D}.$$

In fact, the both sides have the same boundary values $G_\alpha^\nu f|_{\partial D}$ and satisfy the same equation: $(\alpha - W)u = f$ in D.

Thus, applying the operator L to the both sides of formula (4.11), we obtain that

$$LG_\alpha^\nu f = \overline{LH_\alpha}\,(G_\alpha^\nu f|_{\partial D}) + \overline{LG_\alpha^0}f.$$

Since the operators $-\overline{LH_\alpha}^{-1}$ and $\overline{LG_\alpha^0}$ are non-negative, it follows that

$$\left(-\overline{LH_\alpha}^{-1}\right)(LG_\alpha^\nu f) = -G_\alpha^\nu f|_{\partial D} + \left(-\overline{LH_\alpha}^{-1}\right)\left(\overline{LG_\alpha^0}f\right)$$
$$\geq -G_\alpha^\nu f|_{\partial D} \quad \text{on } \partial D.$$

Therefore, by the non-negativity of H_α and G_α^0, we find that

$$G_\alpha f = G_\alpha^\nu f + H_\alpha\left(-\overline{LH_\alpha}^{-1}\,(LG_\alpha^\nu f)\right)$$
$$\geq G_\alpha^\nu f - H_\alpha\,(G_\alpha^\nu f|_{\partial D})$$
$$= G_\alpha^0 f$$
$$\geq 0 \quad \text{on } \bar{D}.$$

ii) Next we prove the boundedness of G_α on the space $C_0(\bar{D}\backslash M)$ with norm

(4.12) $$\|G_\alpha\| \leq \frac{1}{\alpha}\,,\ \alpha > 0.$$

To do so, it suffices to show that

(4.12′) $$f \in C_0(\bar{D}\backslash M)\,,\ f \geq 0 \text{ on } \bar{D} \implies \alpha G_\alpha f \leq \max_{\bar{D}} f \text{ on } \bar{D},$$

since G_α is non-negative on the space $C(\bar{D})$.

We remark (cf. formula (4.3′)) that

$$LG_\alpha^\nu f = \mu\, L_\nu G_\alpha^\nu f + (\mu + \delta - 1)\,(G_\alpha^\nu f|_{\partial D}) - \delta\,((\alpha G_\alpha^\nu f - f)\,|_{\partial D})$$
$$= (\mu + \delta - 1)\,(G_\alpha^\nu f|_{\partial D}) - \delta\,((\alpha G_\alpha^\nu f - f)\,|_{\partial D})\,,$$

so that

(4.5′) $$G_\alpha f = G_\alpha^\nu f - H_\alpha\left(\overline{LH_\alpha}^{-1}\,(LG_\alpha^\nu f)\right)$$
$$= G_\alpha^\nu f + H_\alpha\left(-\overline{LH_\alpha}^{-1}\,((\mu + \delta - 1)G_\alpha^\nu f|_{\partial D})\right)$$
$$+ H_\alpha\left(-\overline{LH_\alpha}^{-1}\,(-\delta(\alpha G_\alpha^\nu f - f)|_{\partial D})\right).$$

Hence we find that the operator G_α is a bounded operator, for each $\alpha > 0$.

Therefore, it suffices to prove the following:

(4.12″) $$f \in C_0(\bar{D}\backslash M) \cap C^\infty(\bar{D})\,,\ f \geq 0 \text{ on } \bar{D} \implies \alpha G_\alpha f \leq \max_{\bar{D}} f \text{ on } \bar{D},$$

since the space $C_0(\bar{D}\backslash M) \cap C^\infty(\bar{D})$ is dense in $C_0(\bar{D}\backslash M)$.

First, applying Theorem A.2 in Appendix with $W - \alpha$ to the function $G_\alpha f$, we obtain that

$$(4.13) \qquad \max_{\bar{D}} G_\alpha f \le \max\left\{ \sup_D \frac{f}{\alpha - W1} \,,\; \max_{\partial D} G_\alpha f \right\}$$

$$\le \max\left\{ \frac{1}{\alpha} \max_{\bar{D}} f \,,\; \max_{\partial D} G_\alpha f \right\}.$$

Now let x' be an arbitrary point of ∂D at which the function $G_\alpha f$ takes its *positive* maximum:

$$G_\alpha f(x') = \max_{\partial D} G_\alpha f = \max_{\bar{D}} G_\alpha f > 0.$$

We remark that

$$\mu(x') + \delta(x') > 0,$$

since $G_\alpha f = 0$ on $M = \{x' \in \partial D; \mu(x') = \delta(x') = 0\}$. Further we have

$$(4.14) \qquad\qquad\qquad\qquad \delta(x') > 0.$$

In fact, if $\delta(x') = 0$ and so $\mu(x') > 0$, then it follows from an application of the boundary point lemma (Lemma A.3) that

$$\mu(x') \frac{\partial}{\partial \mathbf{n}} \left(G_\alpha f\right)(x') < 0.$$

Hence we have

$$
\begin{aligned}
0 &= LG_\alpha f(x') \\
&= \mu(x') L_\nu G_\alpha f(x') + (\mu(x') - 1)G_\alpha f(x') \\
&= \mu(x') \left(\sum_{i,j=1}^{N-1} \bar{\alpha}^{ij}(x') \frac{\partial^2}{\partial x_i \partial x_j} (G_\alpha f)(x') \right) + (\mu(x') - 1)G_\alpha f(x') \\
&\quad + \mu(x') \left(\bar{\eta}(x') + \int_{\partial D} \bar{r}(x', y')[1 - \tau(x', y')]dy' \right. \\
&\quad + \left. \int_D \bar{t}(x', y)[1 - \tau(x', y)]dy \right) G_\alpha f(x') \\
&\quad + \mu(x') \left(\int_{\partial D} \bar{r}(x', y')[G_\alpha f(y') - G_\alpha f(x')]dy' \right. \\
&\quad + \left. \int_D \bar{t}(x', y)[G_\alpha f(y) - G_\alpha f(x')]dy \right) \\
&\quad + \mu(x') \frac{\partial}{\partial \mathbf{n}} (G_\alpha f)(x') \\
&\le \mu(x') \frac{\partial}{\partial \mathbf{n}} (G_\alpha f)(x') \\
&< 0.
\end{aligned}
$$

This contradiction proves assertion (4.14).

If $\delta(x') > 0$, then we have

$$
\begin{aligned}
0 &= LG_\alpha f(x') \\
&= \mu(x') \left(\sum_{i,j=1}^{N-1} \bar{\alpha}^{ij}(x') \frac{\partial^2}{\partial x_i \partial x_j} (G_\alpha f)(x') \right) + (\mu(x') + \delta(x') - 1) G_\alpha f(x') \\
&\quad + \mu(x') \left(\bar{\eta}(x') + \int_{\partial D} \bar{r}(x', y')[1 - \tau(x', y')] dy' \right. \\
&\quad \left. + \int_D \bar{t}(x', y)[1 - \tau(x', y)] dy \right) G_\alpha f(x') \\
&\quad + \mu(x') \left(\int_{\partial D} \bar{r}(x', y')[G_\alpha f(y') - G_\alpha f(x')] dy' \right. \\
&\quad \left. + \int_D \bar{t}(x', y)[G_\alpha f(y) - G_\alpha f(x')] dy \right) \\
&\quad + \mu(x') \frac{\partial}{\partial \mathbf{n}} (G_\alpha f)(x') - \delta(x')(\alpha G_\alpha f(x') - f(x')),
\end{aligned}
$$

and hence

$$
\begin{aligned}
\alpha G_\alpha & f(x') - f(x') \\
&= \frac{1}{\delta(x')} \left[\mu(x') \left(\sum_{i,j=1}^{N-1} \bar{\alpha}^{ij}(x') \frac{\partial^2}{\partial x_i \partial x_j} (G_\alpha f)(x') \right) \right. \\
&\quad + \mu(x') \frac{\partial}{\partial \mathbf{n}} (G_\alpha f)(x') + (\mu(x') + \delta(x') - 1) G_\alpha f(x') \\
&\quad + \mu(x') \left(\bar{\eta}(x') + \int_{\partial D} \bar{r}(x', y')[1 - \tau(x', y')] dy' \right. \\
&\quad \left. + \int_D \bar{t}(x', y)[1 - \tau(x', y)] dy \right) G_\alpha f(x') \\
&\quad + \mu(x') \left(\int_{\partial D} \bar{r}(x', y')[G_\alpha f(y') - G_\alpha f(x')] dy' \right. \\
&\quad \left. \left. + \int_D \bar{t}(x', y)[G_\alpha f(y) - G_\alpha f(x')] dy \right) \right] \\
&\leq 0.
\end{aligned}
$$

This implies that

$$
(4.15) \qquad \max_{\partial D} G_\alpha f = G_\alpha f(x') \leq \frac{1}{\alpha} f(x') \leq \frac{1}{\alpha} \max_{\bar{D}} f.
$$

Therefore, combining estimates (4.13) and (4.15), we obtain that

$$
\max_{\bar{D}} G_\alpha f \leq \frac{1}{\alpha} \max_{\bar{D}} f.
$$

58 KAZUAKI TAIRA

This proves assertion (4.12″) and hence assertion (4.12).

iii) Finally we prove the density of $D(\mathfrak{A})$ in the space $C_0(\bar{D}\backslash M)$: In view of formula (4.9), it suffices to show that

$$(4.16) \qquad \lim_{\alpha\to+\infty} \|\alpha G_\alpha f - f\| = 0, \quad f \in C_0(\bar{D}\backslash M) \cap C^\infty(\bar{D}).$$

We recall (cf. formula (4.5′)) that

$$(4.17) \qquad \alpha G_\alpha f - f = \alpha G_\alpha^\nu f - f - \alpha H_\alpha\left(\overline{LH_\alpha}^{-1}(LG_\alpha^\nu f)\right)$$
$$= (\alpha G_\alpha^\nu f - f) + H_\alpha\left(\overline{LH_\alpha}^{-1}(\alpha(1-\mu-\delta)G_\alpha^\nu f|_{\partial D})\right)$$
$$+ H_\alpha\left(\overline{LH_\alpha}^{-1}(\alpha\delta(\alpha G_\alpha^\nu f - f)|_{\partial D})\right).$$

We estimate the three terms of the second equality in formula (4.17).

iii-1) First, applying Theorem 1 to the boundary condition L_ν (with $\mu = 1$), we find from assertion (3.22) that the first term tends to zero:

$$(4.18) \qquad \lim_{\alpha\to+\infty} \|\alpha G_\alpha^\nu f - f\| = 0.$$

iii-2) To estimate the second term, we remark that

$$H_\alpha\left(\overline{LH_\alpha}^{-1}(\alpha(1-\mu-\delta)G_\alpha^\nu f|_{\partial D})\right)$$
$$= H_\alpha\left(\overline{LH_\alpha}^{-1}((1-\mu-\delta)f|_{\partial D})\right)$$
$$+ H_\alpha\left(\overline{LH_\alpha}^{-1}((1-\mu-\delta)(\alpha G_\alpha^\nu f - f)|_{\partial D})\right).$$

But we have by assertion (4.8)

$$(4.19) \qquad \left\|H_\alpha\left(\overline{LH_\alpha}^{-1}((1-\mu-\delta)(\alpha G_\alpha^\nu f - f)|_{\partial D})\right)\right\|$$
$$\leq \left\|-\overline{LH_\alpha}^{-1}\right\| \cdot \|(1-\mu-\delta)(\alpha G_\alpha^\nu f - f)|_{\partial D}\|$$
$$\leq \frac{1}{k_\alpha}\|(1-\mu-\delta)(\alpha G_\alpha^\nu f - f)|_{\partial D}\|$$
$$\leq \frac{1}{k_1}\|\alpha G_\alpha^\nu f - f\| \longrightarrow 0 \quad \text{as } \alpha\to+\infty.$$

Here we have used the fact:

$$k_1 = -\sup_{x'\in\partial D} LH_1 1(x') \leq k_\alpha = -\sup_{x'\in\partial D} LH_\alpha 1(x') \quad \text{for all } \alpha\geq 1.$$

Thus we are reduced to the study of the term

$$H_\alpha\left(\overline{LH_\alpha}^{-1}((1-\mu-\delta)f|_{\partial D})\right).$$

Now, for any given $\varepsilon > 0$, one can find a function $h \in C^\infty(\partial D)$ such that

$$\begin{cases} h = 0 \text{ near } M = \{x' \in \partial D \,;\, \mu(x') = \delta(x') = 0\}, \\ \|(1 - \mu - \delta)f|_{\partial D} - h\| < \varepsilon. \end{cases}$$

Then we have for all $\alpha \geq 1$

$$(4.20) \qquad \left\| H_\alpha \left(\overline{LH_\alpha}^{-1} \left((1 - \mu - \delta)f|_{\partial D} \right) \right) - H_\alpha \left(\overline{LH_\alpha}^{-1} h \right) \right\|$$
$$\leq \left\| -\overline{LH_\alpha}^{-1} \right\| \cdot \|(1 - \mu - \delta)f|_{\partial D} - h\|$$
$$\leq \frac{\varepsilon}{k_\alpha}$$
$$\leq \frac{\varepsilon}{k_1} \, .$$

Furthermore, one can find a function $\theta \in C_0^\infty(\partial D)$ such that

$$\begin{cases} \theta = 1 & \text{near } M, \\ (1 - \theta)h = h & \text{on } \partial D. \end{cases}$$

Then we have

$$h(x') = (1 - \theta(x'))\, h(x')$$
$$= (-LH_\alpha 1(x')) \left(\frac{1 - \theta(x')}{-LH_\alpha 1(x')} \right) h(x')$$
$$\leq \left[\sup_{x' \in \partial D} \left(\frac{1 - \theta(x')}{-LH_\alpha 1(x')} \right) \right] \|h\| \left(-LH_\alpha 1(x') \right).$$

Since the operator $-\overline{LH_\alpha}^{-1}$ is non-negative on the space $C(\partial D)$, it follows that

$$-\overline{LH_\alpha}^{-1} h \leq \sup_{x' \in \partial D} \left(\frac{1 - \theta(x')}{-LH_\alpha 1(x')} \right) \cdot \|h\| \quad \text{on } \partial D,$$

and hence

$$(4.21) \qquad \left\| H_\alpha \left(\overline{LH_\alpha}^{-1} h \right) \right\| \leq \left\| -\overline{LH_\alpha}^{-1} h \right\| \leq \sup_{x' \in \partial D} \left(\frac{1 - \theta(x')}{-LH_\alpha 1(x')} \right) \cdot \|h\|.$$

But there exists a constant $c_0 > 0$ such that

$$0 \leq \frac{1 - \theta(x')}{\mu(x') + \delta(x')} \leq c_0, \quad x' \in \partial D,$$

so that

$$\frac{1 - \theta(x')}{-LH_\alpha 1(x')} \leq \frac{1 - \theta(x')}{\mu(x') \left(-\frac{\partial}{\partial \mathbf{n}} (H_\alpha 1(x')) \right) + (1 - \mu(x') - \delta(x')) + \alpha \delta(x')}$$

$$\leq \left(\frac{1-\theta(x')}{\mu(x')+\delta(x')}\right)\frac{1}{\min\left\{\inf_{x'\in\partial D}\left(-\frac{\partial}{\partial\mathbf{n}}\left(H_\alpha 1(x')\right)\right),\alpha\right\}}$$

$$\leq c_0 \frac{1}{\min\left\{\inf_{x'\in\partial D}\left(-\frac{\partial}{\partial\mathbf{n}}\left(H_\alpha 1(x')\right)\right),\alpha\right\}}.$$

In view of Lemma 3.19, this implies that

$$\lim_{\alpha\to+\infty}\left[\sup_{x'\in\partial D}\left(\frac{1-\theta(x')}{-LH_\alpha 1(x')}\right)\right]=0.$$

Summing up, we obtain from inequalities (4.20) and (4.21) that

$$\limsup_{\alpha\to+\infty}\left\|H_\alpha\left(\overline{LH_\alpha}^{-1}\left((1-\mu-\delta)f|_{\partial D}\right)\right)\right\|$$

$$\leq \limsup_{\alpha\to+\infty}\left[\left\|H_\alpha\left(\overline{LH_\alpha}^{-1}h\right)\right\|\right.$$

$$\left.+\left\|H_\alpha\left(\overline{LH_\alpha}^{-1}\left((1-\mu-\delta)f|_{\partial D}\right)\right)-H_\alpha\left(\overline{LH_\alpha}^{-1}h\right)\right\|\right]$$

$$\leq \lim_{\alpha\to+\infty}\left[\sup_{x'\in\partial D}\left(\frac{1-\theta(x')}{-LH_\alpha 1(x')}\right)\right]\|h\|+\frac{\varepsilon}{k_1}$$

$$\leq \frac{\varepsilon}{k_1}.$$

Since ε is arbitrary, this proves that

(4.22) $$\lim_{\alpha\to+\infty}\left\|H_\alpha\left(\overline{LH_\alpha}^{-1}\left((1-\mu-\delta)f|_{\partial D}\right)\right)\right\|=0.$$

Therefore, combining assertions (4.19) and (4.22), we find that the second term in formula (4.17) also tends to zero:

$$\lim_{\alpha\to+\infty}\left\|H_\alpha\left(\overline{LH_\alpha}^{-1}\left(\alpha(1-\mu-\delta)G_\alpha^\nu f|_{\partial D}\right)\right)\right\|=0.$$

iii-3) To estimate the third term, we remark that

(4.23) $$\left\|-\overline{LH_\alpha}^{-1}(\alpha\delta)\right\|\leq 1.$$

In fact, we have

$$\alpha\delta(x')=\left(\frac{\alpha\delta(x')}{-LH_\alpha 1(x')}\right)(-LH_\alpha 1(x'))$$

$$\leq \left[\sup_{x'\in\partial D}\left(\frac{\alpha\delta(x')}{-LH_\alpha 1(x')}\right)\right](-LH_\alpha 1(x')),\quad x'\in\partial D,$$

and hence

(4.24) $$\left(-\overline{LH_\alpha}^{-1}\right)(\alpha\delta)\leq \sup_{x'\in\partial D}\left(\frac{\alpha\delta(x')}{-LH_\alpha 1(x')}\right)\quad\text{on }\partial D,$$

since the operator $-\overline{LH_\alpha}^{-1}$ is non-negative on the space $C(\partial D)$. But we find that

$$\frac{\alpha\delta(x')}{-LH_\alpha 1(x')} \le \frac{\alpha\delta(x')}{\mu(x')\left(-\frac{\partial}{\partial\mathbf{n}}\left(H_\alpha 1(x')\right)\right) + (1 - \mu(x') - \delta(x')) + \alpha\delta(x')} \le 1,$$

so that

$$(4.25) \qquad\qquad \sup_{x'\in\partial D}\left(\frac{\alpha\delta(x')}{-LH_\alpha 1(x')}\right) \le 1.$$

Thus, assertion (4.23) follows immediately from inequalities (4.24) and (4.25).

Therefore, combining assertions (4.18) and (4.23), we obtain that the third term tends to zero:

$$\left\|H_\alpha\left(\left(\overline{LH_\alpha}^{-1}\right)(\alpha\delta\left(\alpha G_\alpha^\nu f - f\right)|_{\partial D})\right)\right\|$$
$$\le \left\|\left(-\overline{LH_\alpha}^{-1}\right)(\alpha\delta\left(\alpha G_\alpha^\nu f - f\right)|_{\partial D})\right\|$$
$$\le \left\|-\overline{LH_\alpha}^{-1}(\alpha\delta)\right\|\left\|(\alpha G_\alpha^\nu f - f)|_{\partial D}\right\|$$
$$\le \|\alpha G_\alpha^\nu f - f\| \longrightarrow 0 \quad \text{as } \alpha \to +\infty.$$

This completes the proof of assertion (4.16) and hence that of assertion (iii).

6) Summing up, we have proved that the operator $\mathfrak{A}$, defined by formula (4.4), satisfies conditions (a) through (d) in Theorem 1.3. Hence, in view of assertion (4.2), it follows from an application of part (ii) of the same theorem that the operator $\mathfrak{A}$ is the infinitesimal generator of some Feller semigroup $\{T_t\}_{t\ge 0}$ on $\bar{D}\backslash M$.

The proof of Theorem 4.1 and hence that of Theorem 2 is now complete.

APPENDIX

THE MAXIMUM PRINCIPLE

Let D be a bounded domain of Euclidean space $\mathbf{R}^N$, with boundary ∂D, and let W be a second-order *degenerate* elliptic Waldenfels operator with real coefficients such that

$$Wu(x) = \sum_{i,j=1}^{N} a^{ij}(x)\frac{\partial^2 u}{\partial x_i \partial x_j}(x) + \sum_{i=1}^{N} b^i(x)\frac{\partial u}{\partial x_i}(x) + c(x)u(x)$$

$$+ \int_D s(x,y)\left[u(y) - \sigma(x,y)\left(u(x) + \sum_{j=1}^{N}(y_j - x_j)\frac{\partial u}{\partial x_j}(x)\right)\right]dy,$$

where:

1) $a^{ij} \in C(\mathbf{R}^N)$, $a^{ij} = a^{ji}$ and

$$\sum_{i,j=1}^{N} a^{ij}(x)\xi_i\xi_j \geq 0, \ x \in \mathbf{R}^N, \ \xi = (\xi_1, \cdots, \xi_N) \in \mathbf{R}^N.$$

2) $b^i \in C(\mathbf{R}^N)$, $1 \leq i \leq N$.

3) $c \in C(\mathbf{R}^N)$ and $c \leq 0$ in D.

4) The integral kernel $s(x,y)$ is the distribution kernel of a properly supported, pseudo-differential operator $S \in L_{1,0}^{2-\kappa}(\mathbf{R}^N)$, $\kappa > 0$, which has the transmission property with respect to the boundary ∂D, and $s(x,y) \geq 0$ off the diagonal $\{(x,x); x \in \mathbf{R}^N\}$ in $\mathbf{R}^N \times \mathbf{R}^N$. The measure dy is the Lebesgue measure on $\mathbf{R}^N$.

5) The function $\sigma(x,y)$ is a C^∞ function on $\bar{D} \times \bar{D}$ such that $\sigma(x,y) = 1$ in a neighborhood of the diagonal $\{(x,x); x \in \bar{D}\}$ in $\bar{D} \times \bar{D}$.

6) $W1(x) = c(x) + \int_D s(x,y)[1 - \sigma(x,y)]dy \leq 0$ in D.

First we have the following result:

Theorem A.1 (The weak maximum principle). *Assume that a function $u \in C(\bar{D}) \cap C^2(D)$ satisfies either*

$$Wu \geq 0 \ \text{and} \ W1 < 0 \ \text{in} \ D$$

or

$$Wu > 0 \ \text{and} \ W1 \leq 0 \ \text{in} \ D.$$

Then the function u may take its positive maximum only on the boundary ∂D.

As an application of the weak maximum principle, we can obtain a pointwise estimate for solutions of the inhomogeneous equation $Wu = f$:

Theorem A.2. *Assume that*

$$W1 < 0 \quad \text{on} \quad \bar{D} = D \cup \partial D.$$

Then we have for all $u \in C(\bar{D}) \cap C^2(D)$

$$\max_{\bar{D}} |u| \leq \max\left\{ \left(\frac{1}{\min_{\bar{D}}(-W1)}\right) \sup_D |Wu|, \max_{\partial D} |u| \right\}.$$

Now assume that D is a *domain of class* C^2, that is, each point of the boundary ∂D has a neighborhood in which ∂D is the graph of a C^2 function of $N-1$ of the variables $x_1, \cdots, x_N$. We consider a function $u \in C(\bar{D}) \cap C^2(D)$ which satisfies the condition

$$Wu \geq 0 \quad \text{in } D,$$

and study the interior normal derivative $\frac{\partial u}{\partial \mathbf{n}}$ at a point where the function u takes its non-negative maximum.

We define a subset Σ_3 of the boundary ∂D by

$$\Sigma_3 = \left\{ x' \in \partial D \; ; \; \sum_{i,j=1}^N a^{ij}(x')n_i n_j > 0 \right\},$$

where $\mathbf{n} = (n_1, \cdots, n_N)$. In other words, the set Σ_3 is the set of non-characteristic points with respect to the operator W. It is easy to see that Σ_3 is invariant under C^2 diffeomorphisms which preserve normal vectors.

The boundary point lemma reads as follows:

Lemma A.3 (The boundary point lemma). *Let D be a domain of class C^2. Assume that a function $u \in C(\bar{D}) \cap C^2(D)$ satisfies the condition*

$$Wu \geq 0 \quad \text{in } D,$$

and that there exists a point x'_0 of the set Σ_3 such that

$$\begin{cases} u(x'_0) = \max_{x \in \bar{D}} u(x) \geq 0, \\ u(x) < u(x'_0), \quad x \in D. \end{cases}$$

Then the interior normal derivative $\frac{\partial u}{\partial \mathbf{n}}(x'_0)$ of u at x'_0, if it exists, satisfies the condition

$$\frac{\partial u}{\partial \mathbf{n}}(x'_0) < 0.$$

For a proof of Theorems A.1 and A.2 and Lemma A.3, the reader might refer to Bony-Courrège-Priouret [1], Oleĭnik-Radkevič [7] and Taira [12].

REFERENCES

1. J.-M. Bony, P. Courrège et P. Priouret, *Semi-groupes de Feller sur une variété à bord compacte et problèmes aux limites intégro-différentiels du second ordre donnant lieu au principe du maximum*, Ann. Inst. Fourier (Grenoble) **18** (1968), 369-521.

2. L. Boutet de Monvel, *Boundary problems for pseudo-differential operators*, Acta Math. **126** (1971), 11-51.

3. C. Cancelier, *Problèmes aux limites pseudo-différentiels donnant lieu au principe du maximum*, Comm. P.D.E. **11** (1986), 1677-1726.

4. R.R. Coifman et Y. Meyer, *Au-delà des opérateurs pseudo-différentiels*, Astérisque No. 57, Soc. Math. France, Paris, 1978.

5. L. Hörmander, *Pseudodifferential operators and non-elliptic boundary problems*, Ann. of Math. **83** (1966), 129-209.

6. H. Kumano-go, *Pseudo-differential operators*, MIT Press, Cambridge, Mass., 1981.

7. O.A. Oleĭnik and E.V. Radkevič, *Second order equations with nonnegative characteristic form*, (in Russian), Itogi Nauki, Moscow, 1971; *English translation*, Amer. Math. Soc., Providence, Rhode Island and Plenum Press, New York, 1973.

8. S. Rempel and B.-W. Schulze, *Index theory of elliptic boundary problems*, Akademie-Verlag, Berlin, 1982.

9. K. Sato and T. Ueno, *Multi-dimensional diffusion and the Markov process on the boundary*, J. Math. Kyoto Univ. **14** (1965), 529-605.

10. R.T. Seeley, *Singular integrals and boundary value problems*, Amer. J. Math. **88** (1966), 781-809.

11. K. Taira, *Sur l'existence de processus de diffusion*, Ann. Inst. Fourier (Grenoble) **29** (1979), 99-126.

12. _____, *Diffusion processes and partial differential equations*, Academic Press, Boston San Diego London Tokyo, 1988.

13. _____, *Elliptic boundary value problems, analytic semigroups and Markov processes*, Lecture Notes in Math. Vol. 1499, Springer-Verlag, Berlin Heidelberg New York Tokyo, 1991.

14. S. Takanobu and S. Watanabe, *On the existence and uniqueness of diffusion processes with Wentzell's boundary conditions*, J. Math. Kyoto Univ. **28** (1988), 71-80.

15. M. Taylor, *Pseudodifferential operators*, Princeton Univ. Press, Princeton, 1981.

16. H. Triebel, *Interpolation theory, function spaces, differential operators*, North-Holland Publishing Company, Amsterdam New York Oxford, 1978.

17. A.D. Wentzell (Ventcel'), *On boundary conditions for multidimensional diffusion processes* (in Russian), Teoriya Veroyat. i ee Primen. **4** (1959), 172-185; English translation in Theory Prob. and its Appl. **4** (1959), 164-177.

18. K. Yosida, *Functional analysis*, Springer-Verlag, Berlin Heidelberg New York, 1965.

INSTITUTE OF MATHEMATICS, UNIVERSITY OF TSUKUBA, TSUKUBA 305, JAPAN

Editorial Information

To be published in the *Memoirs*, a paper must be correct, new, nontrivial, and significant. Further, it must be well written and of interest to a substantial number of mathematicians. Piecemeal results, such as an inconclusive step toward an unproved major theorem or a minor variation on a known result, are in general not acceptable for publication. *Transactions* Editors shall solicit and encourage publication of worthy papers. Papers appearing in *Memoirs* are generally longer than those appearing in *Transactions* with which it shares an editorial committee.

As of July 1, 1992, the backlog for this journal was approximately 8 volumes. This estimate is the result of dividing the number of manuscripts for this journal in the Providence office that have not yet gone to the printer on the above date by the average number of monographs per volume over the previous twelve months. (There are 6 volumes per year, each containing about 3 or 4 numbers.)

A Copyright Transfer Agreement is required before a paper will be published in this journal. By submitting a paper to this journal, authors certify that the manuscript has not been submitted to nor is it under consideration for publication by another journal, conference proceedings, or similar publication.

Information for Authors

Memoirs are printed by photo-offset from camera copy fully prepared by the author. This means that the finished book will look exactly like the copy submitted.

The paper must contain a *descriptive title* and an *abstract* that summarizes the article in language suitable for workers in the general field (algebra, analysis, etc.). The *descriptive title* should be short, but informative; useless or vague phrases such as "some remarks about" or "concerning" should be avoided. The *abstract* should be at least one complete sentence, and at most 300 words. Included with the footnotes to the paper, there should be the 1991 *Mathematics Subject Classification* representing the primary and secondary subjects of the article. This may be followed by a list of *key words and phrases* describing the subject matter of the article and taken from it. A list of the numbers may be found in the annual index of *Mathematical Reviews*, published with the December issue starting in 1990, as well as from the electronic service e-MATH [**telnet e-MATH.ams.com** (or **telnet 130.44.1.100**). Login and password are **e-math**]. For journal abbreviations used in bibliographies, see the list of serials in the latest *Mathematical Reviews* annual index. When the manuscript is submitted, authors should supply the editor with electronic addresses if available. These will be printed after the postal address at the end of each article.

Electronically-prepared manuscripts. The AMS encourages submission of electronically-prepared manuscripts in $\mathcal{A}_{\mathcal{M}}\mathcal{S}$-TEX or $\mathcal{A}_{\mathcal{M}}\mathcal{S}$-LaTEX. To this end, the Society has prepared "preprint" style files, specifically the amsppt style of $\mathcal{A}_{\mathcal{M}}\mathcal{S}$-TEX and the amsart style of $\mathcal{A}_{\mathcal{M}}\mathcal{S}$-LaTEX, which will simplify the work of authors and of the production staff. Those authors who make use of these style files from the beginning of the writing process will further reduce their own effort.

Guidelines for Preparing Electronic Manuscripts provide additional assistance and are available for use with either $\mathcal{A}_{\mathcal{M}}\mathcal{S}$-TeX or $\mathcal{A}_{\mathcal{M}}\mathcal{S}$-LaTeX. Authors with FTP access may obtain these *Guidelines* from the Society's Internet node e-MATH.ams.com (130.44.1.100). For those without FTP access they can be obtained free of charge from the e-mail address guide-elec@math.ams.com (Internet) or from the Publications Department, P. O. Box 6248, Providence, RI 02940-6248. When requesting *Guidelines* please specify which version you want.

Electronic manuscripts should be sent to the Providence office only after the paper has been accepted for publication. Please send electronically prepared manuscript files via e-mail to pub-submit@math.ams.com (Internet) or on diskettes to the Publications Department address listed above. When submitting electronic manuscripts please be sure to include a message indicating in which publication the paper has been accepted.

For papers not prepared electronically, model paper may be obtained free of charge from the Editorial Department at the address below.

Two copies of the paper should be sent directly to the appropriate Editor and the author should keep one copy. At that time authors should indicate if the paper has been prepared using $\mathcal{A}_{\mathcal{M}}\mathcal{S}$-TeX or $\mathcal{A}_{\mathcal{M}}\mathcal{S}$-LaTeX. The *Guide for Authors of Memoirs* gives detailed information on preparing papers for *Memoirs* and may be obtained free of charge from AMS, Editorial Department, P. O. Box 6248, Providence, RI 02940-6248. The *Manual for Authors of Mathematical Papers* should be consulted for symbols and style conventions. The *Manual* may be obtained free of charge from the e-mail address cust-serv@math.ams.com or from the Customer Services Department, at the address above.

Any inquiries concerning a paper that has been accepted for publication should be sent directly to the Editorial Department, American Mathematical Society, P. O. Box 6248, Providence, RI 02940-6248.

Recent Titles in This Series

(*Continued from the front of this publication*)

(See the AMS catalog for earlier titles)